Circuits and Electronics

Circuits and Electronics
Hands-on Learning with Analog Discovery

John Okyere Attia

CRC Press
Taylor & Francis Group
Boca Raton London New York

CRC Press is an imprint of the
Taylor & Francis Group, an **informa** business

CRC Press
Taylor & Francis Group
6000 Broken Sound Parkway NW, Suite 300
Boca Raton, FL 33487-2742

Printed on acid-free paper

International Standard Book Number-13: 978-1-138-29732-6 (Hardback)

Visit the Taylor & Francis Web site at
http://www.taylorandfrancis.com

and the CRC Press Web site at
http://www.crcpress.com

To my brothers and sisters (Augustine, Kwaku, Francisca, Esther, Adwoa, Anthony, Clement, and Samuel) for their ceaseless love. May the sacred bond amongst us grow stronger and stronger each day.

Contents

Examples

Preface

Hands-on learning is a method whereby students actively perform some activities in order to learn from them. Hands-on learning may involve students in building, testing, observing, and reflecting on completed work. Hands-on learning is said to be especially useful to students who learn by doing. In addition, by reflecting on work that has been done, additional ideas might crop up in the learner's brain that may be investigated to achieve deeper learning. A saying attributed to Confucius sums up the advantages of hands-on learning: "I hear and I forget. I see and I remember. I do and I understand." One tool for hands-on learning is the Analog Discovery board.

The Analog Discovery board was developed by Digilent in conjunction with Analog Devices, Inc. The Analog Discovery board communicates to a computer through a USB interface. The operating software, WaveForms, provides virtual instrument functions. The virtual instruments include a two-channel voltmeter, a two-channel oscilloscope, a two-channel waveform generator, a 16-channel logic analyzer, and two fixed dc power supplies, Network Analyzer, and Spectrum Analyzer. The virtual instruments are equivalent to more expensive desktop instruments when the board is connected to a laptop computer through its USB port.

Traditionally, hands-on learning had taken place in laboratories with well-equipped instruments. With the Analog Discovery board, experiments can be done anywhere and anytime. The Analog Discovery board enables students to quickly test real-world functional circuits with their own personal computers. This book is not a laboratory manual for performing experiments in circuit and electronics. On the contrary, the book discusses concepts, theory, and laws in circuits and electronics with examples solved using the Analog Discovery board. Virtual Instruments of the Analog Discovery, such as scope, the signal generator, Network Analyzer, Spectrum Analyzer, and the voltmeter, are described and used throughout the book.

The book provides instructions on building circuits on breadboards, connecting the Analog Discovery wires to the circuit under test, and making electrical measurements. Various measurement techniques are described and used in this book. They include impedance measurements, complex power measurements, frequency response measurements, power spectrum measurements, current versus voltage characteristic measurements of diodes, bipolar junction transistors, and MOSFETs. The end-of-chapter problems have been provided to give the student additional exercises for hands-on learning and experimentation, and to give comparisons between measured results and those obtained from theoretical calculations.

The goals of writing this book are (1) to provide the reader with an introduction to the use of the virtual instruments of the Analog Discovery board; (2) to provide the reader with simple, step-by-step instructions in the use of the Analog Discovery board for solving problems in circuits and electronics; (3) to assist the reader in connecting theory to practice in the fields of circuits and electronics with the use of the Analog Discovery board; and (4) to provide the student with problems that allow them to reflect upon the discrepancies between the measured and the theoretically calculated results in circuits and electronics.

This book is unique. It is the first time a book has been written that covers an introduction to the virtual instruments of the Analog Discovery board concisely with examples on the use of virtual instruments for solving problems in circuits and electronics. In addition, it is the first time a book has been written that connects the theory and practice in circuits and electronics through the use of the Analog Discovery board.

Audience

The book can be used by students, professional engineers, and technicians who want a basic introduction to the use of the Analog Discovery board, which can be found in Chapter 1 of the book. Chapters 2 to 4 discuss hands-on learning in circuit analysis. Basic circuit laws, transient analysis, impedances, power calculations and frequency response are discussed in the three chapters. The chapters may be useful to students who want to do hands-on learning in circuit analysis with the Analog Discovery board. Chapters 5, 6, and 7 are for electrical and electrical engineering students and professionals who want to use Analog Discovery to explore the characteristics of operational amplifiers, diodes, bipolar junction transistors, and metal-oxide-semiconductor field effect transistors.

Organization

Chapter 1 is a basic introduction to the flywire and virtual instruments of the Analog Discovery board. Problems have been solved by performing basic measurements on circuits through the use of the virtual instruments of the Analog Discovery board.

Chapters 2, 3, and 4 cover hands-on learning in circuit analysis. Chapter 2 discusses basic circuit laws and theorems. Ohm's Law, Kirchhoff's Voltage Law, Kirchhoff's Current Law, homogeneity property, Thevenin's and

Norton's Theorems are discussed with examples done with the Analog Discovery board to allow comparisons between theoretically calculated results and those obtained through measurements. The transient response of RC, RL, and RLC circuits are discussed in Chapter 3. The examples in Chapter 3 provide hands-on learning with the Analog Discovery board and the measurement results are compared with those obtained from the theory. Chapter 4 discusses topics in sinusoidal steady state analysis, including impedance, power calculations, and frequency response. Some measurement techniques discussed in the chapter include those for determining impedance, complex power, and magnitude and phase characteristics of circuits.

Chapters 5, 6, and 7 cover hands-on learning in analog electronics. Amplifier configurations, integrators, differentiators, and active filters are discussed in Chapter 5. The virtual instruments of the Analog Discovery board are used to obtain voltage, add two signals, convert square waves to triangular waveforms, and obtain the frequency responses of active filters. Chapter 6 discusses various diode circuits. The Analog Discovery board is used to obtain current versus voltage characteristics of diodes and to explore the characteristics of half-wave and full-wave rectifier circuits. Bipolar junction and field effect transistors are discussed in Chapter 7. The Analog Discovery board has been used to obtain current versus voltage characteristics of transistors, operating points, and frequency responses of bipolar junction and field effect transistor amplifiers.

End-of-chapter problems are available to assist in the hands-on learning by students, to compare the results obtained by measurements and those obtained from theory.

A bibliography appears at the end of the book.

Analog Discovery Board and Analog Discovery 2

This book describes the features and instruments of the Analog Discovery board. All the examples have been solved with the Analog Discovery board. A newer version of the Analog Discovery board, Analog Discovery 2, uses the same color-coded wire harness as that of the Analog Discovery board. However, the Analog Discovery 2 uses the WaveForms 2015 software, which has more virtual instruments than the WaveForms software of the Analog Discovery board. The steps for building the circuits in this book will not change whether one uses the Analog Discovery board or the Analog Discovery 2. However, the procedures in activating the virtual instruments of Analog Discovery provided in this book will not be the same as those steps needed to activate the instruments of Analog Discovery 2.

MATLAB® is a registered trademark of The MathWorks, Inc. For product information, please contact:

The MathWorks, Inc.
3 Apple Hill Drive
Natick, MA 01760-2098 USA
Tel: 508-647-7000
Fax: 508-647-7001
E-mail: info@mathworks.com
Web: www.mathworks.com

Acknowledgments

I thank the individuals who introduced me to the Analog Discovery board: Prof. Kenneth Connor of Rensselaer Polytechnic Institute, Prof. Mohammed Chouikha of Howard University, and Prof. Yacob Astake of Morgan State University. In addition, I am grateful to the group members of the "Experimental Centric Based Engineering Curriculum for HBCUs" for their discussions on the practical applications of the Analog Discovery board. Furthermore, I am grateful to Prof. Robert Bowman of Rochester Institute of Technology for his advice on using the Analog Discovery board to obtain current-voltage characteristics of transistors.

Author

Dr. John Okyere Attia is professor of electrical and computer engineering at Prairie View A&M University. He was the department head of the Electrical and Computer Engineering Department from 1997 to 2013. Dr. Attia earned a PhD in electrical engineering from the University of Houston, an M.S. from the University of Toronto, and a BS from the Kwame Nkrumah University of Science and Technology.

Dr. Attia teaches graduate and undergraduate courses in electrical and computer engineering in the fields of electronics, circuit analysis, instrumentation systems, digital signal processing, and VLSI design. His research interests include hands-on learning, power electronics for microgrid, innovative electronic circuit designs for radiation environments, and radiation testing methodologies of electronic devices and circuits.

Dr. Attia has worked on projects funded by the National Science Foundation, NASA, Texas Higher Education Coordinating Board, and the Texas Workforce Commission. Dr. Attia has more than 80 technical publications. He has published in refereed journal papers and refereed conference papers. In addition, he is the author of four books published by the CRC Press. His previous books are *Electronics and Circuit Analysis Using MATLAB, 2nd Edition*, and *PSPICE and MATLAB for Electronics: An Integrated Approach, 2nd Edition*.

Dr. Attia has received several honors. Amongst them are the Most Outstanding Senior in the Graduating Class, the Outstanding Teacher Award, Exemplary University Achiever, and the University Leader Award. Dr. Attia is a member of the following honor societies: Sigma Xi, Tau Beta Pi, Kappa Alpha Kappa, and Eta Kappa Nu. In addition, Dr. Attia is an ABET program evaluator for electrical and computer engineering programs. Furthermore, he is a senior member of the Institute of Electrical and Electronics Engineers and a registered Professional Engineer in the State of Texas.

1

Virtual Instruments of the Analog Discovery Board

1.1 The Analog Discovery Board

The Analog Discovery (AD) platform consists of an Analog Discovery board, shown in Figure 1.1, and virtual instrument software called WaveForms, which is described in Section 1.3.

The Analog Discovery board has a wire harness with pins for connecting to differential oscilloscopes, waveform generators, power supplies, and a logic analyzer. The pin-out of the wire harness is shown in Figure 1.2.

The flywire (from www.digilentinc.com) has leads for connecting the Analog Discovery board for signal generation and display. In addition, there are leads for power supply outputs and digital I/O signals. Table 1.1 shows the leads and their color representations.

1.2 The WaveForms Software

The Digilent WaveForms software is a suite of virtual instruments available to you on your personal computer. The software can be downloaded from the Digilent website. The WaveForms software connects to the Analog Discovery board and provides an easy-to-use graphical interface for virtual instruments. The WaveForms software can be used to produce analog and digital signals and can acquire, analyze, and store signals. When the WaveForms software is launched, the opening window shown in Figure 1.3 appears.

The Analog Discovery virtual instruments consist of the following (for analog instruments):

1. Oscilloscope (two-channel scope)
2. Arbitrary waveform generator (two generators)

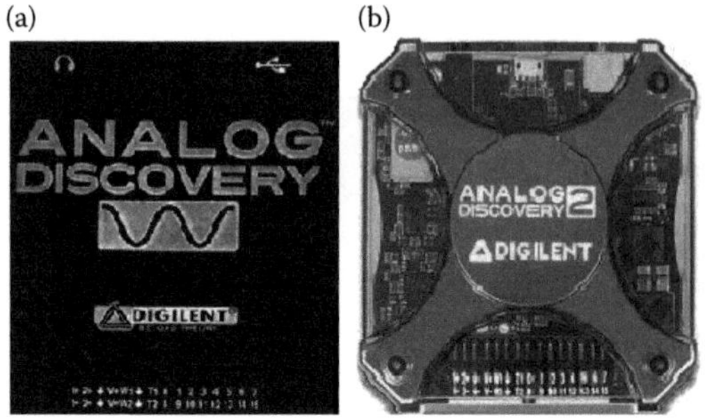

FIGURE 1.1
(a) Analog Discovery and (b) Analog Discovery 2.

3. Power supplies (5 and –5 V sources)
4. Voltmeter
5. Network analyzer
6. Spectrum analyzer.

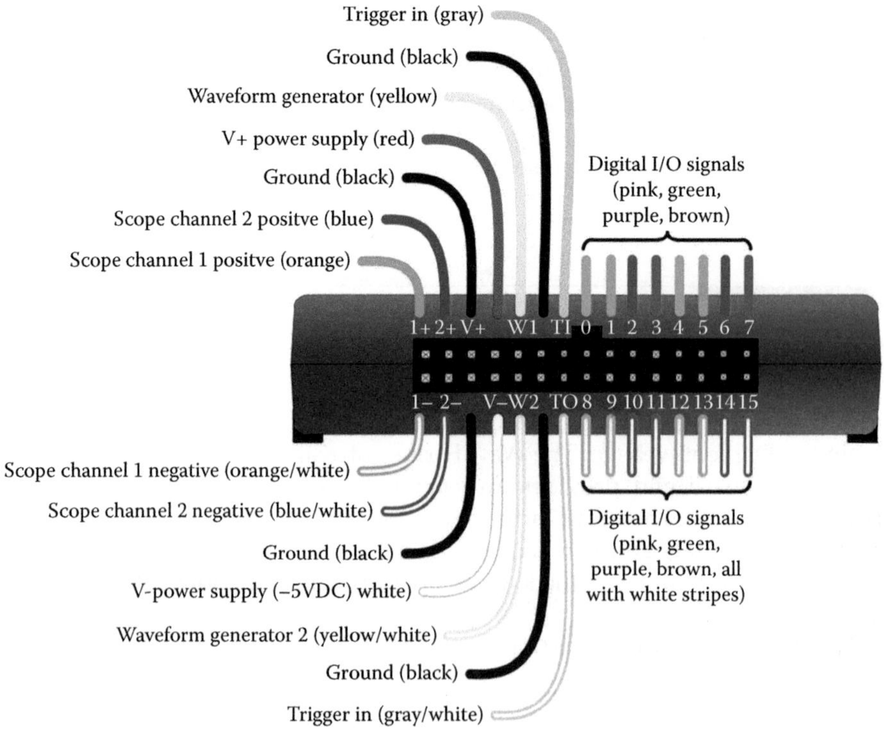

FIGURE 1.2
Pin-Out of Fly-wire of Analog Discovery board (see Table 1.1 for color representations).

TABLE 1.1

Analog Discovery Board Wire Harness

Function	Color
Scope channel 1 (positive)	Orange
Scope channel 1 (negative)	Orange-white
Scope channel 2 (positive)	Blue
Scope channel 2 (negative)	Blue-white
Ground	Black
V+ power supply (5 V dc)	Red
V– power supply (–5 V dc)	White
Arbitrary waveform generator 1	Yellow
Arbitrary waveform generator 2	Yellow-white
16 Digital I/O signals	Various colors

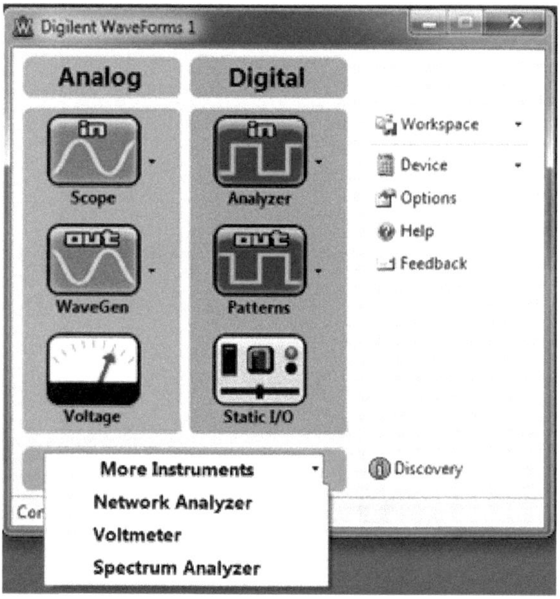

FIGURE 1.3
Main window of WaveForms software showing the virtual instruments.

The Analog Discovery virtual instruments consist of the following (for digital instruments):

1. Logic analyzer
2. Digital pattern generator
3. Static I/O.

This book only discusses the applications of the analog virtual instruments of the Analog Discovery board. The virtual instruments are described in the following sections.

1.3 Power Supplies

The Analog Discovery board has two fixed power supplies: (1) +5 V and (2) −5 V. If one clicks on the voltage button in the Main WaveForms window, the power supply window shown in Figure 1.4 appears. To turn on the power supply, click the Power Is OFF button. Another click on the same button turns the power supply off. One can click the V+ button to turn on the +5 V power supply. The −5 V source can be turned on by clicking the V− power supply button.

1.4 Voltmeters

There are two voltmeters, found under the More Instruments tab. The voltmeters can measure dc voltage, True rms, and ac rms voltages. The voltmeter

FIGURE 1.4
Power supplies screen of Analog Discovery board.

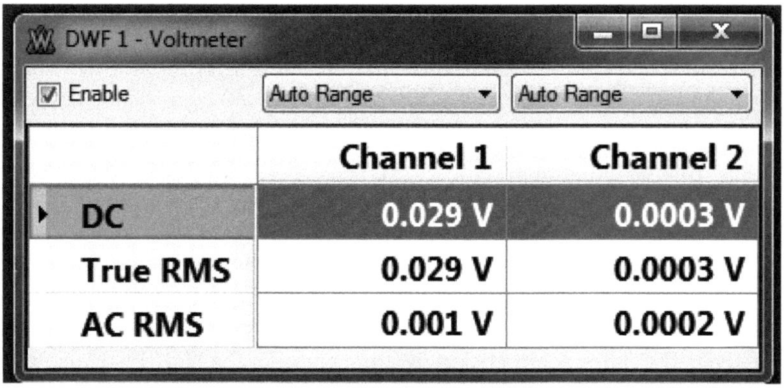

FIGURE 1.5
Voltmeter screen of Analog Discovery board.

window appears by clicking the Voltmeter button in More Instruments. The voltmeter screen is shown in Figure 1.5. Example 1.1 shows how to measure voltages by using the Analog Discovery voltmeter.

Example 1.1 Use of the Analog Discovery Power Supply and Voltmeter

For the circuit shown in Figure 1.6, use the Analog Discovery board to find the voltage across the 100 kΩ resistor.

***Solution*:**
Build the Circuit:
Use a breadboard to build the circuit shown in Figure 1.6. Connect a 5 V voltage source (red wire) to node 1 of the circuit. Also, connect scope channel 1 positive (orange wire) to node 1. In addition, connect the scope channel 1 negative (orange-white wire) to node 0. Furthermore, connect the ground (black wire) to node 0 of the circuit. Moreover, connect scope

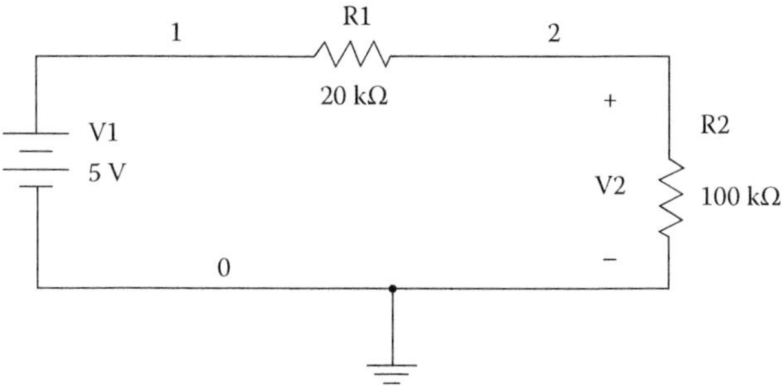

FIGURE 1.6
Resistive circuit.

channel 2 positive (blue) to node 2 and scope channel 2 negative (blue-white) to node 0. The scope channel 1 measures the voltage of the source, and the scope channel 2 measures the voltage across resistor R2. The circuit built on a breadboard is shown in Figure 1.7.

Activation of the Power Supply:
Click Voltage in the main WaveForms screen → Click Power to turn on the voltage sources → Click V+ to turn on the 5 V source → Upon completion of the measurements, click Power again to turn it off.

(*Note*: → indicates the sequence of action steps needed to use the virtual instruments of the Analog Discovery board.)

Activation of Voltmeter:
Activate the voltmeter by clicking Voltmeter under the More Instruments tab. Click Enable to start reading the voltage of the power supply source and the voltage across resistor R2.

The scope channel 1 measures the voltage of the power supply (nodes 1 and 0), and the scope scope channel 2 shows the voltage across R2 (nodes 2 and 0). The voltmeter readings are shown in Figure 1.8. Using the voltage divider rule, the voltage across the resistor R2 is calculated as

$$V2 = (100\ K) * (5) / 120\ K = 4.16\ V$$

The measured value is 4.088 V, which is about a 1.7% deviation from the calculated value.

The deviation of the measured value from the calculated value can be attributed to the fact that the resistors used for breadboarding have a tolerance of 10%, and to the fact that the source voltage is not exactly 5 V, as shown in channel 1 of the voltmeter reading.

FIGURE 1.7
Circuit of Figure 1.6 built on a breadboard.

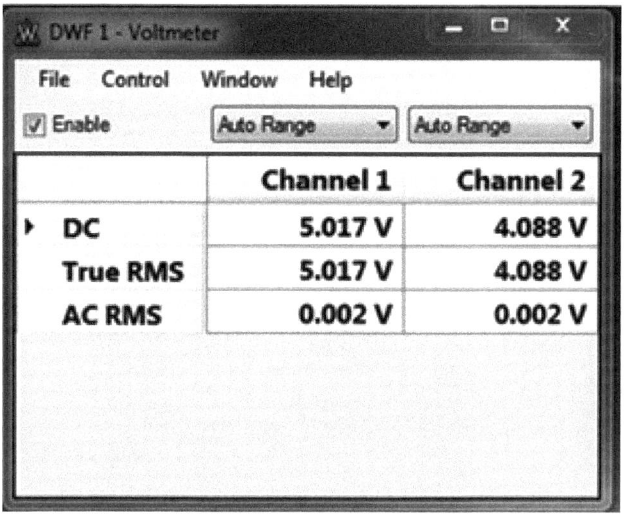

FIGURE 1.8
Voltmeter readings of the voltage of the power supply (channel 1) and voltage across R2 (channel 2).

1.5 Arbitrary Waveform Generators

The arbitrary waveform generator can be used to generate basic, advanced, and user-defined waveforms.

1.5.1 Basic Waveforms

The arbitrary waveform generator showing the basic signals that can be generated is shown in Figure 1.9. The signals that can be generated consist of the waveforms shown in Table 1.2.

If the dc source is selected, the voltage of the source can be varied from –5 to +5 V. However, if one selects sinusoid, triangular, or other waveform signals shown in Table 1.2, the frequency, amplitude, and offset can be changed.

1.5.2 Sweep Waveforms

The available sweep signals are shown in Table 1.3. For the signals shown in Table 1.3, the frequency may be swept from a starting value to an ending value within a specified duration of time. A screenshot of the sweep waveform is shown in Figure 1.10. The amplitude of the sweep waveform may be set from values of –5 to +5 V. The duration of the sweep can be specified.

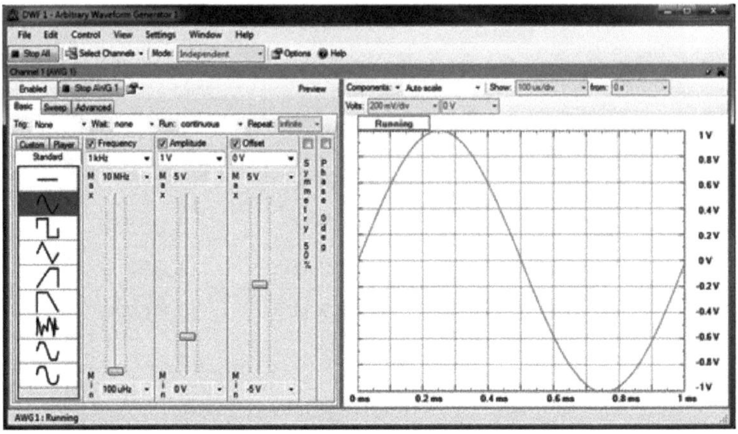

FIGURE 1.9
Arbitrary waveform generator showing basic signals.

TABLE 1.2
Basic Signals Available from Arbitrary Waveform Generator Window

DC
Sinusoid
Square
Triangular
Ramp-up
Ramp-down
Noise
Trapezium
Sine-Power

TABLE 1.3
Sweep Waveforms Available from Arbitrary Waveform Generator Window

Sinusoid
Square
Triangular
Ramp-up
Ramp-down
Noise

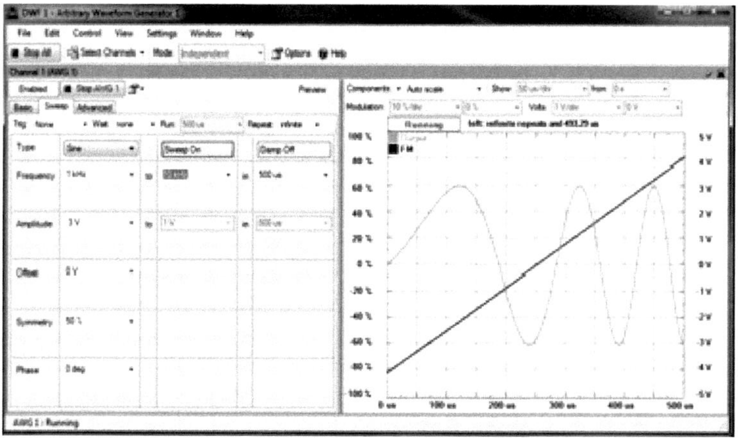

FIGURE 1.10
Arbitrary waveform generator showing a sweep signal.

1.5.3 Advanced Waveforms

The available waveforms under the Advanced button are amplitude modulated (AM) and frequency modulated (FM) waveforms. Figure 1.11 shows an example of an amplitude modulated waveform. For the AM wave, the amplitude and frequency of the carrier signal can be selected. In addition, the frequency of the modulating signal and the modulation index can also be selected. For the FM wave (not shown), the amplitude and frequency of the carrier can be selected. In addition, modulating frequency plus the frequency modulation index can be chosen.

There are two arbitrary waveform generators, AWG1 and AWG2. Figure 1.12 shows the two waveform generators, displaying a sine wave in AWG1 and

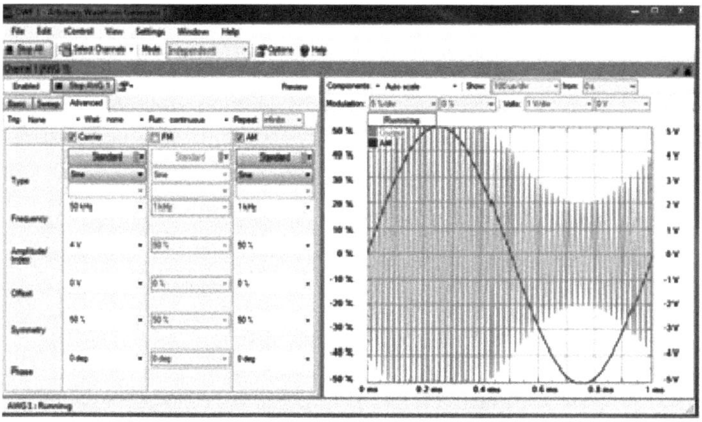

FIGURE 1.11
An amplitude modulated waveform.

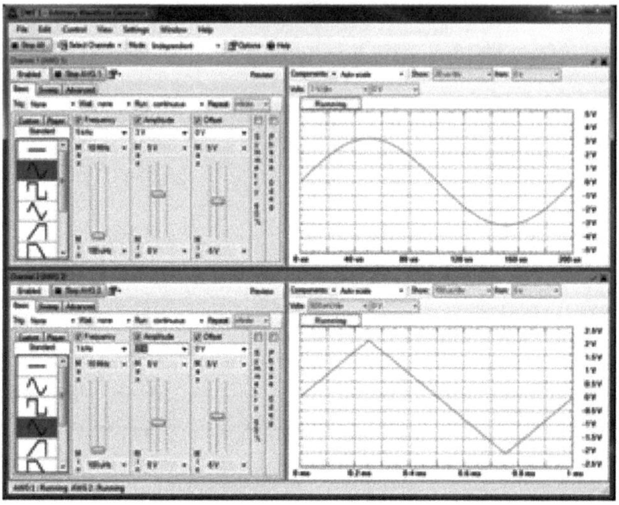

FIGURE 1.12
Arbitrary waveform generators AWG1 and AWG2.

a triangular waveform in AWG2. Example 1.2 shows the use of AWG1 to generate a dc voltage source for a resistor–LED circuit.

Example 1.2 Resistor–LED Circuit

For the resistor–LED circuit shown in Figure 1.13, (1) determine the voltage drop across the LED, and (2) calculate the current flowing through the LED. (Use a red LED.)

Build the Circuit:
Use a breadboard to build the circuit shown in Figure 1.13. Connect AWG1 (yellow wire) to node 1 of the circuit. Also, connect scope channel 1 positive (orange wire) to node 1. Furthermore, connect the scope channel 1 negative (orange-white wire) to node 2. The scope channel 1 measures the voltage across the resistor. In addition, connect the ground (black wire) to node 0 of the circuit. Moreover, connect scope channel 2

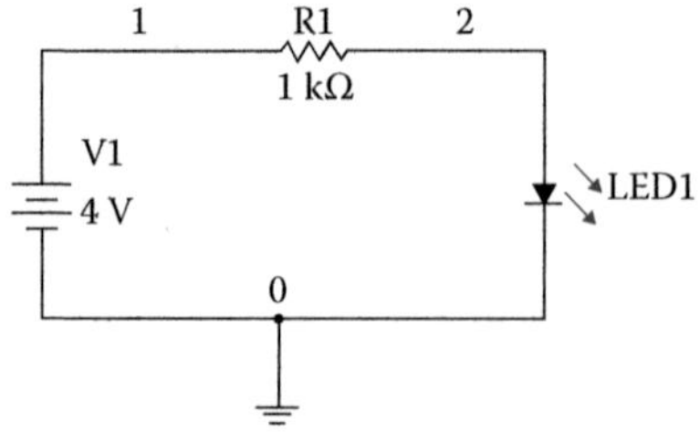

FIGURE 1.13
Resistor–LED circuit.

positive (blue) to node 2 and scope channel 2 negative (blue-white) to node 0. The scope channel 2 shows the voltage across the LED.

Signal Generation:
Click WaveGen in the main WaveForms screen. Go to Basic → dc → Offset (set to 4 V) → Click Run AWG1.

Activation of Voltmeter:
Activate the Voltmeter by clicking Voltmeter under the More Instruments tab. Click Enable to start reading the voltage across the resistor and the LED.

Figure 1.14 shows the resistor–LED circuit built on a breadboard. Figure 1.15 shows the dc voltage generated from Analog Discovery

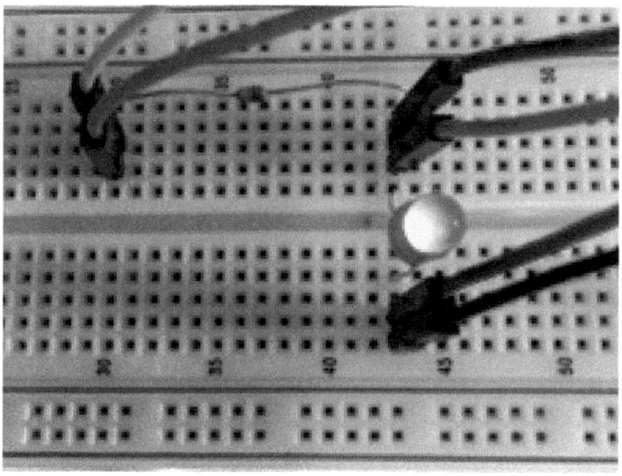

FIGURE 1.14
Resistor–LED circuit built on a breadboard.

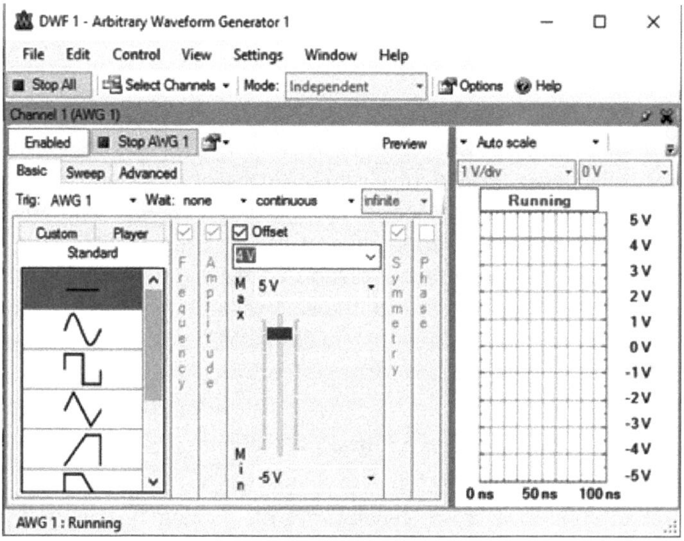

FIGURE 1.15
DC signal generated from Analog Discovery AWG1.

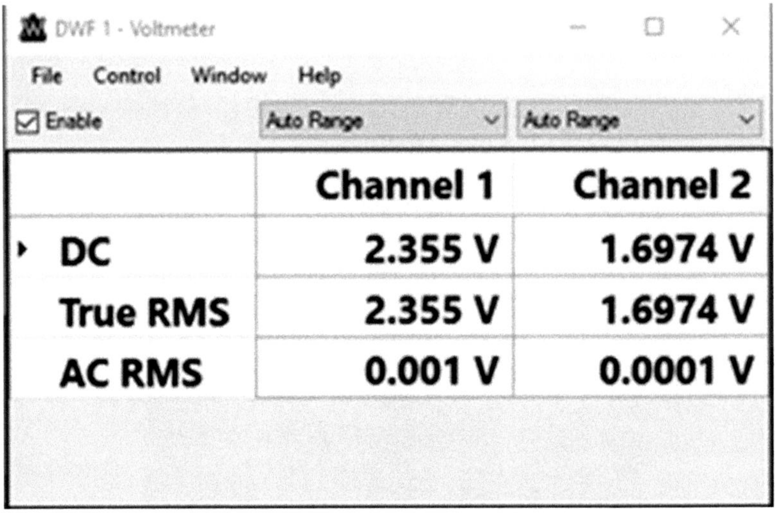

FIGURE 1.16
Voltmeter readings of the voltage across resistor (channel 1) and voltage across LED (channel 2).

AWG1. In addition, Figure 1.16 shows the measured voltages across the resistor and LED. From the voltmeter readings, the voltage across the resistor is 2.355 V, and that across the LED is 1.6974 V. Using Ohm's Law, the current, I, flowing through the LED is given as

$$\text{Current, } I = 2.355 \text{ V}/1 \text{ K}\Omega = 2.355 \text{ mA}$$

1.6 Scope

The Analog Discovery board has two oscilloscopes (scopes) that can be used to display voltages. The scopes have several trigger sources and can have ac/dc coupling. In addition, the scopes can display FFT and XY data. Furthermore, scope data can be exported in multiple formats to other application software, such as Microsoft Excel. Example 1.3 shows how to use the Analog Discovery scope to display a sinusoidal voltage generated by the Analog Discovery arbitrary waveform generators (AWG1 and AWG2).

Example 1.3 Displays of an Arbitrary Waveform Generator Signal on a Scope

Use the arbitrary waveform generator of the Analog Discovery board to produce a sinusoidal signal with amplitude of 3 V and frequency of 2 kHz. Display the generated sinusoidal signal on the Analog Discovery scope.

Solution:

Build the Circuit:

Use a breadboard to build the circuit shown in Figure 1.17. Connect AWG1 (yellow wire) to node 1 of the circuit. Also, connect scope channel 1 positive (orange wire) to node 1. In addition, connect the scope channel 1 negative (orange-white wire) to node 0. Moreover, connect the ground (black wire) to node 0 of the circuit. The picture of the connected circuit is shown in Figure 1.18.

Generation of Sine Wave:

Click WaveGen in the main WaveForms screen. Go to Basic → sine → Frequency (set to 1 kHz) → Amplitude (set to 3 V) → Offset (set to 0 V) → Symmetry (set to 50%) → Click Run AWG1. Figure 1.19 shows the setup of Digilent Analog Discovery WaveGen.

Activation of Scope:

Click the Scope in the main WaveForms screen. Go to Run → Autoset → Measure → Add → Channel 1 Horizontal → Frequency → Add → Channel 1 Vertical → Peak to Peak → Add Selected Measurement → Close Add Instrument. The waveform seen on the scope is shown in Figure 1.20. It can be seen from the scope display that the frequency measured is 1.99 kHz.

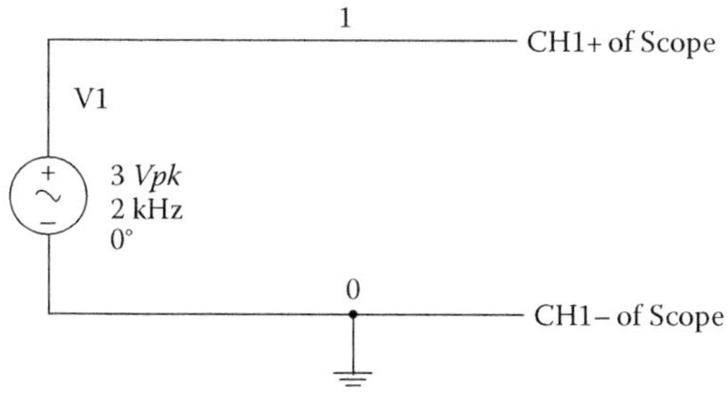

FIGURE 1.17
Signal generator connected to scope.

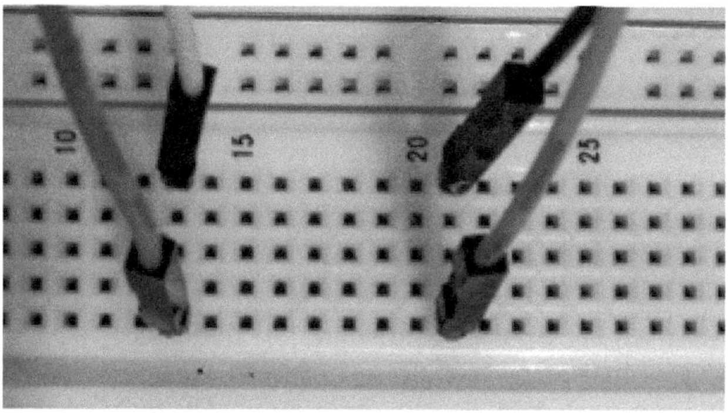

FIGURE 1.18
Picture of the connected wires.

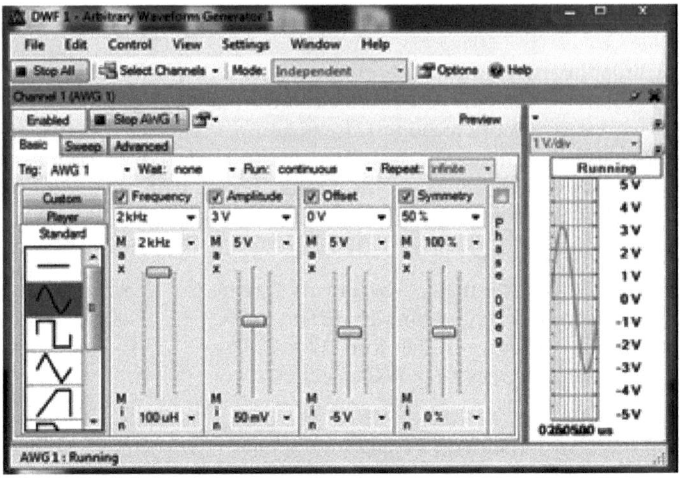

FIGURE 1.19
The setup of arbitrary waveform generator.

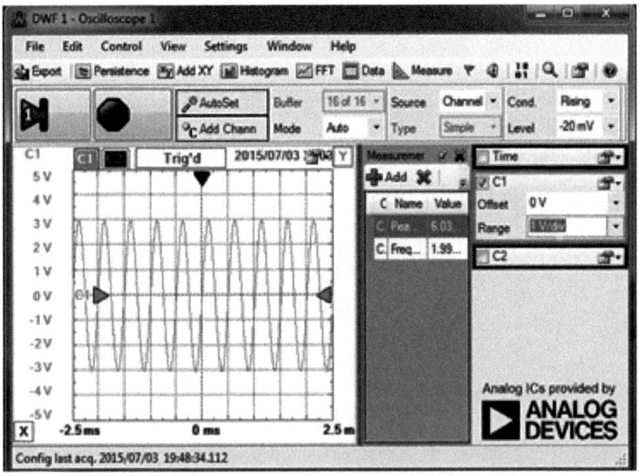

FIGURE 1.20
Scope display of a signal generated by the arbitrary waveform generator.

1.7 The Network Analyzer

The Network Analyzer is used to display the Bode plot (which consists of magnitude-versus-frequency and phase-versus-frequency plots). The Network Analyzer has two main areas (1) the control and (2) the plot areas. The control area contains panels that allow one to adjust the settings for the Bode plot. The plot area contains the Bode plot. The control area functions are given below:

- *Single button*: starts a single analysis.
- *Run/Stop button*: starts repeated analysis or stops.
- *Start*: lets you specify the start frequency.
- *Stop*: lets you specify the stop frequency.
- *AWG Offset*: lets you specify the offset for the generated signal.
- *Amplitude*: lets you specify the amplitude of the generated signal.
- *Steps*: lets you specify the number of steps for the analysis.
- *Max-Gain*: lets you specify the maximum gain of the filter. The gain can be specified independently for each channel under the Scope Channels drop-down menu.
- *Bode Scale*: opens a drop-down menu where the top and range of magnitude and also phase plots can be adjusted.

To display the Bode plot, the AWG1 and the scope channel 1 input are both connected to the input of the circuit under test. The scope channel 2 is connected to the output of the circuit under test. Example 1.4 shows how to use the Network Analyzer to obtain the Bode plot of a circuit.

Example 1.4 Magnitude and Phase Characteristics of an RLC Circuit

For the RLC circuit shown in Figure 1.21, use the Analog Discovery board to determine the magnitude and phase characteristics of the circuit. Use the following component values for the circuit: R1 = 1 kΩ, L1 = 40 mH, and C1 = 0.01 μF.

Solution:
Build the Circuit:
Use a breadboard to build the circuit shown in Figure 1.21. Connect AWG1 (yellow wire) to node 1 of the circuit. Also, connect scope channel 1 positive (orange wire) to node 1. In addition, connect the scope channel 1 negative (orange-white wire) to node 0. Moreover, connect the ground (black wire) to node 0 of the circuit. Furthermore, connect scope

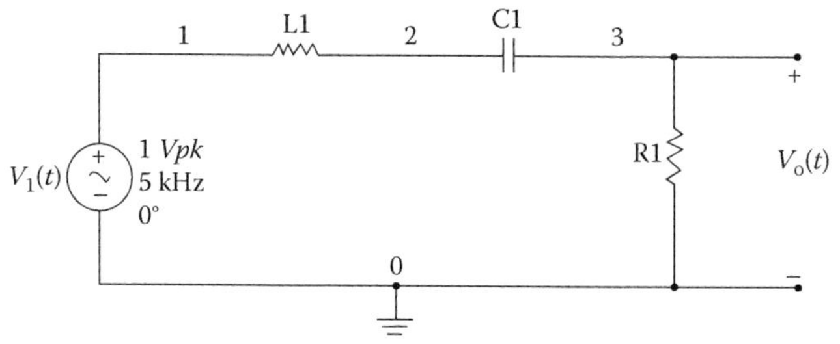

FIGURE 1.21
RLC circuit.

channel 2 positive (blue wire) to node 3, and the scope channel 2 negative (blue-white wire) to node 0. The RLC circuit, built using a breadboard, is shown in Figure 1.22.

Generation of Sine Wave:
Click WaveGen in the main WaveForms screen. Go to Basic → Sinusoid → Frequency (set to 5 kHz) → Amplitude (set to 1 V) → Offset (set to 0 V) → Symmetry (set to 50%) → Click Run AWG1.

Activation of Network Analyzer:
The Network Analyzer uses both oscilloscope channels as the input and output channels. Channel 1 is used for measuring the input signal and channel 2 for measuring the output signal. Open up the Network Analyzer that can be found under the More Instruments tab. Use the following values for setup of the Network Analyzer:

Start Frequency: 1 kHz
Stop Frequency: 50 kHz
Offset: 0 V
Input Signal Amplitude: 1 V
Max Filter Gain: 1X
Bode Scale: Magnitude—Top 5 dB, Range of 40 dB
Phase—Top 90°, Range of 180°
Scope Channel Gain: Channel 1: 1X; Channel 2: 1X

Obtain a single sweep in frequency by clicking Single. This provides a Bode plot representation of the frequency response of the circuit. The Bode plot is shown in Figure 1.23.

FIGURE 1.22
RLC circuit built on a breadboard.

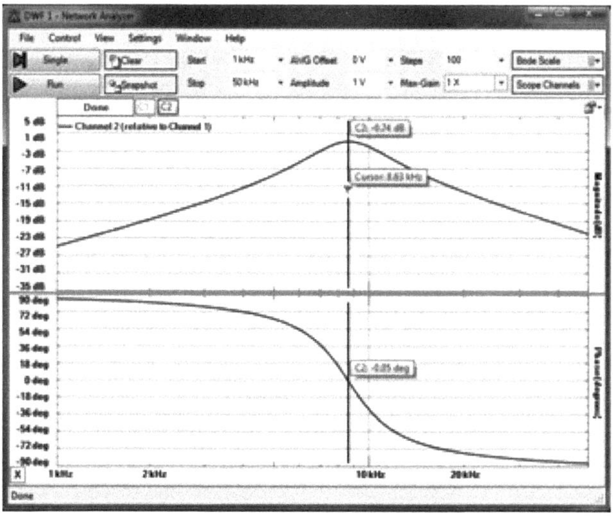

FIGURE 1.23
Magnitude and phase response of the RLC circuit of Figure 1.21.

1.8 Spectrum Analyzer

The Spectrum Analyzer is primarily used to obtain the power spectrum of signals. It measures the magnitude versus frequency and the phase versus frequency of a signal. It can be used to obtain the dominant frequency and the harmonics of a periodic signal. The Spectrum Analyzer uses FFT or Chirp-Z transform to obtain the spectrum of an input signal. It uses the oscilloscope channels to acquire the input signal. Thus, when the Spectrum Analyzer starts sampling, instruments that use the scope channels will stop acquiring signals. The Spectrum Analyzer instrument is found under the More Instruments tab in the Analog Discovery WaveForms software package. The following features are available:

1. Frequency:

   Frequency ranges—select the operation range of the Spectrum Analyzer.

   Center/Span:

   The Center frequency is the midpoint between the stop and start frequencies. The Span parameter states the range between the Start and Stop frequencies.

   Start/Stop:

   Instead of using the Center/Span frequencies, one can specify the Start and Stop frequencies.

Tracking:

Enabling Tracking allows the Spectrum Analyzer to search for the maximum signal level on the screen and makes that frequency the center frequency of the signal. Tracking may be disabled.

2. Level

The following parameters can be changed under Level:

Units:

Units select the amplitude units. They can be:

V_{peak}—relative to 1 V amplitude sine wave.

V_{rms}—relative to 1 V (rms) sine wave.

Ref—the channel input range is configured according to this reference setting.

3. Trace

Up to 10 traces can be added. For each input channel, the window functions can be changed.

4. Window functions

Window functions can be found under the Options drop-down menu. Some window functions that can be selected are: rectangular, triangular, Hamming, Hanning, cosine, and Blackman-Harris.

5. Control menu

The control menu has the following functions:

Single—starts a single acquisition of signal.

Run—starts repeated acquisition of a signal.

Stop—stops signal acquisition.

Example 1.5 shows how to use the Spectrum Analyzer to determine the frequency content of a periodic signal.

Example 1.5 Frequency Spectrum of a Square Wave

Determine the frequency content of a square wave with peak value of 2 V, zero average value, and a periodic frequency of 1 kHz.

Solution:

Build the Circuit:

Use a breadboard to build the circuit shown in Figure 1.24. Connect AWG1 (yellow wire) to node 1 of the circuit. Also, connect scope channel 1 positive (orange wire) to node 1. In addition, connect the scope channel 1 negative (orange-white wire) to node 0. Moreover, connect the ground (black wire) to node 0 of the circuit. A picture of the connected circuit is similar to that shown in Figure 1.18.

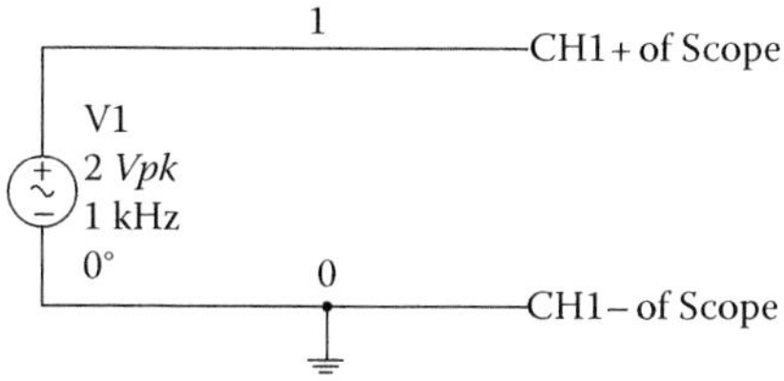

FIGURE 1.24
Connection for obtaining frequency content of a signal.

Generation of Square Wave:
Click WaveGen in the main WaveForms screen. Go to Basic → Square → Frequency (set to 1 kHz) → Amplitude (set to 2 V) → Offset (set to 0 V) → Symmetry (set to 50%) → Click Run AWG1.

Activation of Scope:
Click Scope in the main WaveForms screen. Go to Run → Autoset. The waveform seen on the scope is shown in Figure 1.25.

Activation of Spectrum Analyzer:
The Spectrum Analyzer uses the scope channel as the input. Open up the Spectrum Analyzer (found in the More Instruments tab in the Analog Discovery Waveform software package). Use the following setup values for the Spectrum Analyzer:

Frequency Range: 10 kHz to 24.42 Hz
Frequency
  Center: 5 kHz
  Span: 10 kHz
  Track: Disabled

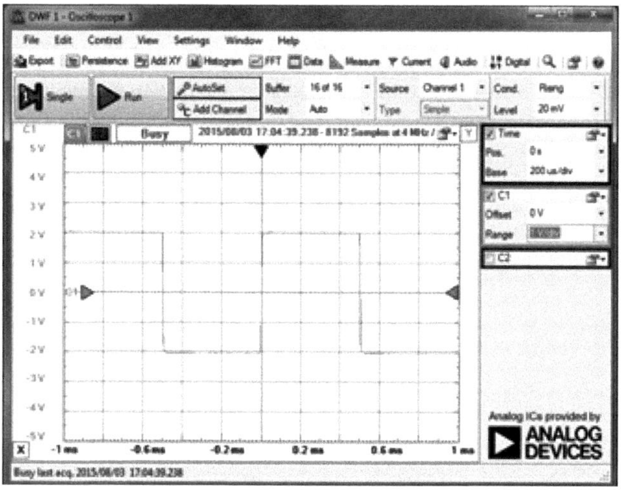

FIGURE 1.25
Square wave displayed on a scope.

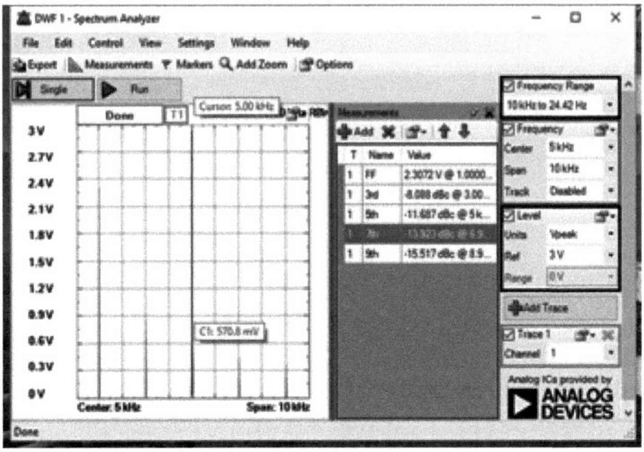

FIGURE 1.26
Frequency spectra of a square wave.

Level
  Units: *V*peak
  Ref: 3 V
Trace 1: Channel 1

Run the Spectrum Analyzer. Measure the frequency of the spectral lines and their magnitude (peak value) by using the cursor. Capture the spectra lines. The spectral lines of the square wave are shown in Figure 1.26.

It can be seen from Figure 1.26 that the spectral lines occur around the following frequencies: 1, 3, 5, 7, and 9 kHz.

PROBLEMS

Problem 1.1 For the circuit shown in Figure P1.1, (1) use the Analog Discovery board to find the voltage across the 1 kΩ resistor. (2) As V1 increases from 1 to 5 V with 1 V increments, record the voltages across R2. (3) What is the relationship between the output and the input voltages?

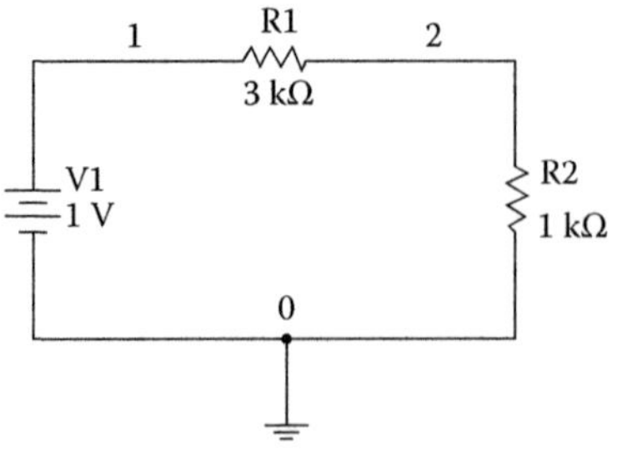

FIGURE P1.1
Resistive circuit.

Problem 1.2 For the resistor–LED circuit shown in Figure P1.2, (1) determine the voltage drop across the LED. (2) Calculate the current flowing through the LED. Use green LED. (3) Compare the voltage drop across the green LED to that obtained for the red LED of Example 1.2.

Problem 1.3 For the circuit shown in Figure P1.3, (1) display the output voltage on a scope. (2) When the frequency of the source voltage is increased to 100 Hz, what is your observation with regard to the LED?

Problem 1.4 For the circuit shown in Figure P1.4, the input square wave signal has a frequency of 100 Hz, the peak-to-peak value of the square wave is

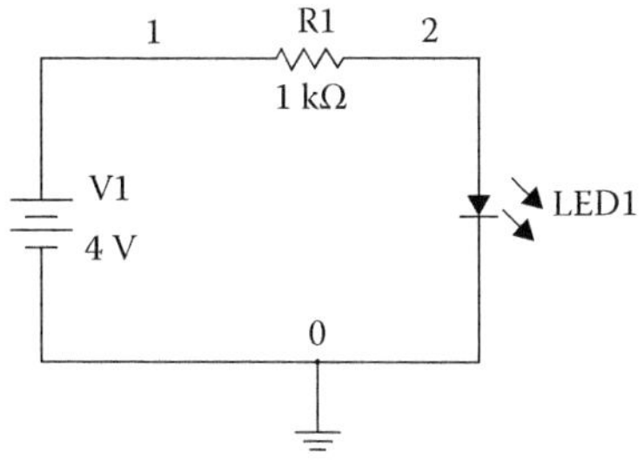

FIGURE P1.2
Resistor–LED circuit.

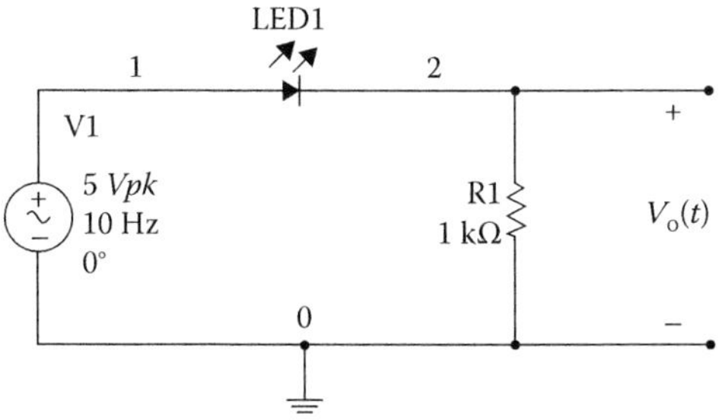

FIGURE P1.3
Half-wave rectifier circuit with LED.

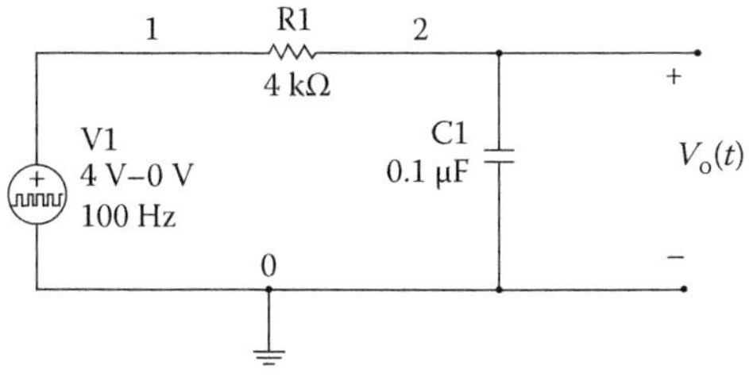

FIGURE P1.4
RC circuit.

2 V with offset voltage of 2 V. (1) Use the Analog Discovery board to obtain the output signal. (2) What happens to the output voltage when the source frequency is changed to 1 kHz?

Problem 1.5 For the RLC circuit shown in Figure P1.5, use the Analog Discovery board to determine the magnitude and phase characteristics at nodes 2 and 0.

Problem 1.6 (1) Determine the frequency content of a triangular waveform with peak value of 2 V, zero average value, and a periodic frequency of 1 kHz. (2) Compare the results of part (1) with those obtained in Example 1.5.

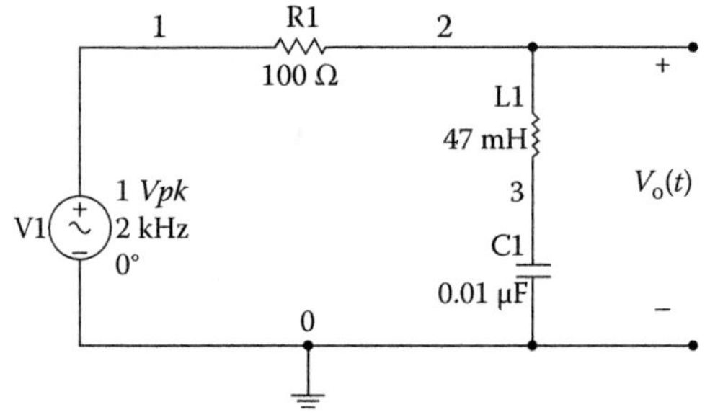

FIGURE P1.5
RLC circuit.

2

Basic Circuit Laws and Theorems

INTRODUCTION In this chapter, the basic circuit laws and theorems will be explored by using the Analog Discovery board. The circuit laws and theorems that will be explored include Ohm's Law, Kirchhoff's Voltage Law, Kirchhoff's Current Law, Thevenin's and Norton's Theorems. In addition, the linearity principle will be discussed.

2.1 Ohm's Law

Ohm's Law states that the voltage, $v(t)$, across a resistor is directly proportional to the current, $i(t)$, flowing through the resistor. Equation 2.1 shows the Ohm's Law relationship

$$v(t) = Ri(t) \tag{2.1}$$

where R is the proportionality constant

Example 2.1 shows how the Analog Discovery board can be used to verify Ohm's Law.

Example 2.1 Ohm's Law

For the circuit shown in Figure 2.1, use the Analog Discovery board to show that the voltage across the resistor is proportional to the current flowing through the resistor. Assume that $R = 1.0\ \text{k}\Omega$.

Solution:

Build the Circuit:

Use a breadboard to build the circuit shown in Figure 2.1. Connect AWG1 (yellow wire) to node 1 of the circuit. Also, connect scope channel 1 positive (orange wire) to node 1. In addition, connect the scope channel 1 negative (orange-white wire) to node 0. Moreover, connect the ground (black wire) to node 0 of the circuit.

Signal Generation:

Click WaveGen in the main WaveForms screen. Go to Basic → dc → Offset (set to 1 V) → Click Run AWG1.

Activation of Scope:

Click the scope in the main WaveForms screen. Go to Run → Autoset → Measure → Add → Channel 1 Vertical → Maximum → Add Selected Measurement → Close Add Instrument.

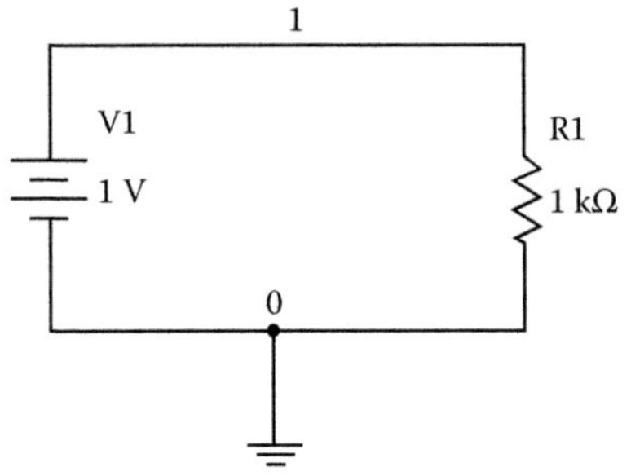

FIGURE 2.1
Resistive circuit.

TABLE 2.1
Voltage and Current of a Resistive Circuit

Voltage (V)	Current (mA)
1.013	1.013
2.011	2.011
3.009	3.009
4.008	4.008
5.009	5.009

To add a Mathematic Channel to measure the current, use the following steps: Go to Control → Add Mathematic Channel → Custom → Enter function → C1/1000 → OK.

To get the value for the current, go to Measure → Add → Math 1 Vertical → Maximum → Add Selected Measurement → Close Add Instrument. By default, the units in the math channel will be volts. To change the units to amperes, click the Setting button → select Units → select A from the drop-down menu.

Increase the dc voltage source from 1 to 5 V with increments of 1 V. Table 2.1 shows the values of the voltage across the resistor and the corresponding current flowing through the resistor.

It can be seen from Table 2.1 that the voltage is proportional to current, thus confirming Ohm's Law.

2.2 Kirchhoff's Voltage Law

The Kirchhoff's Voltage Law (KVL) states that the algebraic sum of all the voltages around a closed loop is zero. Mathematically, KVL can be stated as

$$\sum_{n=1}^{K} V_n = 0 \tag{2.2}$$

where K is the number of voltages in a loop, and V_n is the n-th voltage.

Example 2.2 shows how to use the Analog Discovery board to confirm KVL.

Example 2.2 Kirchhoff's Voltage Law

For the circuit shown in Figure 2.2, use the Analog Discovery board to show that the sum of voltages across the resistors is the same as the voltage of the dc source.

Solution:

Build the Circuit:

Use a breadboard to build the circuit shown in Figure 2.3. Connect AWG1 (yellow wire) to node 1 of the circuit. Also, connect scope channel 1 positive (orange wire) to node 1. In addition, connect the scope channel 1 negative (orange-white wire) to node 0. Moreover, connect the ground (black wire) to node 0 of the circuit. Scope channel 1 measures the voltage of the source. Furthermore, connect scope channel 2 positive (blue) to node 1 and connect scope channel 2 negative (blue-white) to node 2. This measures the voltage across resistor R1.

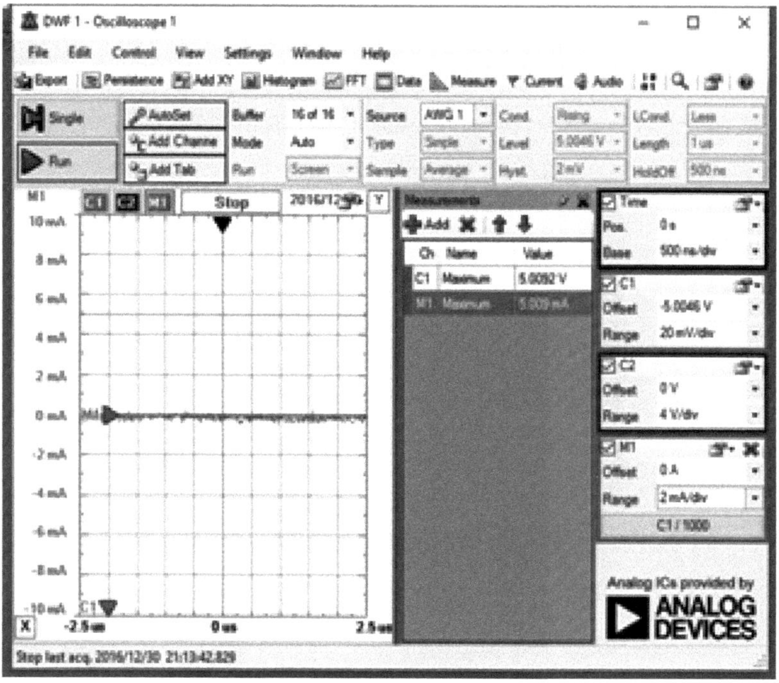

FIGURE 2.2
The source voltage and current of circuit shown in Figure 2.1.

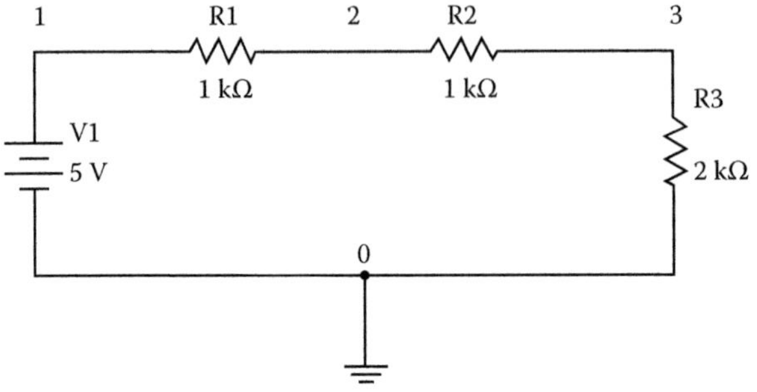

FIGURE 2.3
Voltage source plus three resistors circuit.

Signal Generation:
Click WaveGen in the main WaveForms screen. Go to Basic → dc → Offset (set to 5 V) → Click Run AWG1.

Activation of Voltmeter:
Activate the voltmeter by clicking Voltmeter under the More Instruments tab. Click Enable to start reading the voltage across R1 (nodes 1 and 2).

Use the scope channel 2 to measure the voltage across R2 (nodes 2 and 3). In addition, use the scope channel 2 to measure the voltage across R3 (nodes 3 and 0). Record your measurements. Figure 2.4 shows the voltage across the dc source and the voltage across the resistor R1. Table 2.2 shows the voltages at various nodes of the circuit shown in Figure 2.3.

From Table 2.2, it can be seen that the voltage of the source is almost equal to the sum of the voltages across the resistors, R1, R2, and R3, thus confirming the KVL.

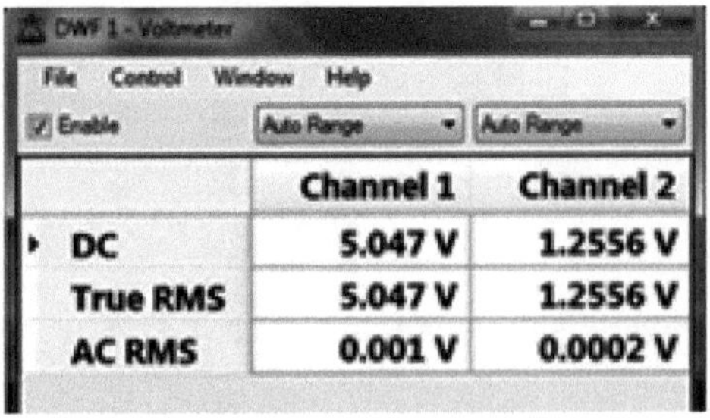

FIGURE 2.4
Voltages across the dc source V1 and resistor R1 measured by the voltmeter.

TABLE 2.2

DC Voltages Measured by Voltmeter of the Analog Discovery Board

Voltage across R1 (V)	Voltage across R2 (V)	Voltage across R3 (V)	Sum of Voltages across R1, R2, and R3	Voltage V1 (V)
1.256	1.255	2.512	5.027	5.047

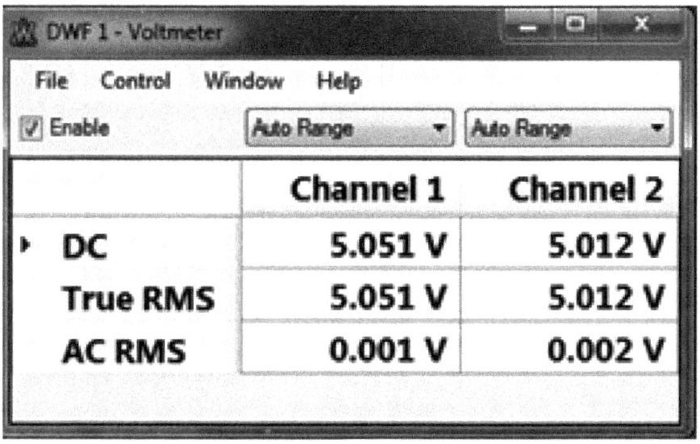

FIGURE 2.5
Source voltage of 5 V measured by the voltmeter.

The dc voltages across the dc source were measured using channel 1 and channel 2 of the voltmeter. It was found that the voltages are not same, although the source voltage was 5 V. This is shown in Figure 2.5.

2.3 Kirchhoff's Current Law

The Kirchhoff's Current Law (KCL) states that the algebraic sum of all the currents at a node is zero. Mathematically, KCL can be stated as

$$\sum_{n=1}^{K} i_n = 0 \tag{2.3}$$

where K is the number of branches connected to a node, and i_n is the n-th current entering or leaving a node.

Example 2.3 shows how to use the Analog Discovery board to confirm KCL.

Example 2.3 Kirchhoff's Current Law

For the circuit shown in Figure 2.6, use the Analog Discovery board to show that the algebraic sum of all the currents at node B is zero.

Solution:
Build the Circuit:
Use a multimeter to measure the resistances of resistors R1, R2, R3, and R4. Use a breadboard to build the circuit shown in Figure 2.6. Connect AWG1 (yellow wire) to node A of the circuit. Also, connect the ground (black wire) to node D of the circuit.

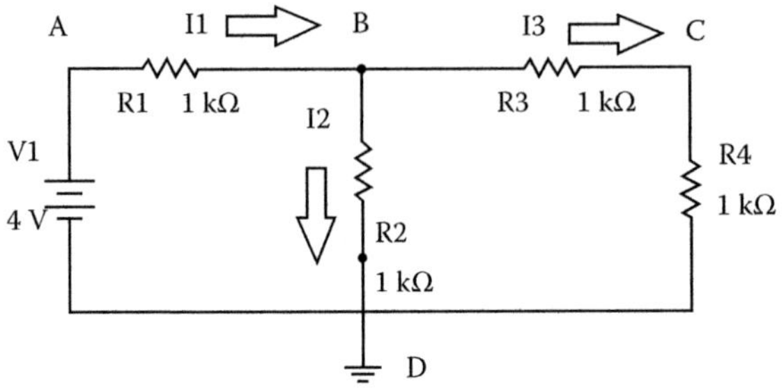

FIGURE 2.6
DC source and resistive circuit.

Signal Generation:
Click WaveGen in the main WaveForms screen. Go to Basic → dc → Offset (set to 4 V) → Click Run AWG1. Figure 2.7 shows voltage generated by arbitrary waveform generator AWG1.

Voltage Measurements:
Use the multimeter to measure the voltage across resistors R1, R2, and R3. Use Ohm's Law to calculate the current flowing through the resistors R1, R2, and R3.

Table 2.3 shows the voltages across the resistors R1, R2, and R3 and the corresponding currents I1, I2, and I3 are shown in Table 2.4. The currents were calculated by using Ohm's Law.

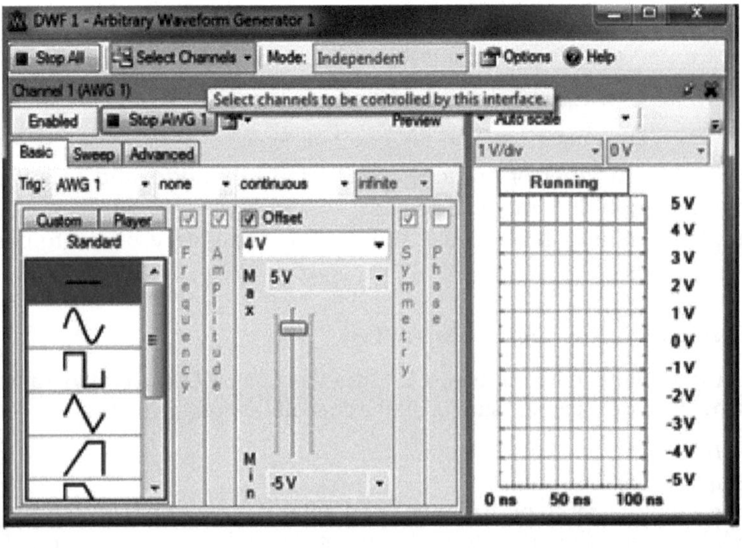

FIGURE 2.7
Signal generated by arbitrary waveform generator AWG1.

TABLE 2.3

Measured Resistances and Voltages

Measured Resistances (Ω)				Measured Voltages (V)			
R1	R2	R3	R4	VR1	VR2	VR3	VR4
984	986	986	991	2.391	1.599	0.797	0.801

TABLE 2.4

Currents in Figure 2.6 Calculated by Using Voltages and Resistances in Table 2.3

Current in mA				
I2 (VR2/R2)	I3=(VR3/R3)	I4 (VR4/R4)	I1 (VR1/R1)	I2+I3
1.62	0.81	0.81	2.43	2.43

The current I4 is the current flowing through resistor R4, and it is in the same direction as current I3 (The current I4 is not shown in Figure 2.6). It can be seen from Table 2.4 that the current I1=I2+I3. In addition, current I3=I4. The results confirm KCL.

2.4 Linearity Property

Linearity property describes a linear relation between cause and effect. The linearity property is the combination of the homogeneity (scaling) property and the additive property. The homogeneity property requires that if an input is multiplied by a constant *K*, the output will also be multiplied by the same constant *K*. The additive property says that the response to a sum of inputs is the sum of the responses to each input when applied separately. The homogeneity property will be explored in Example 2.4.

Example 2.4 The Homogeneity Property of a Linear Circuit

For the circuit shown in Figure 2.8, show that the voltage across the 2 kΩ resistor is linearly proportional to the source voltage.

Solution:

Build the Circuit:

Use a breadboard to build the circuit shown in Figure 2.8. Connect AWG1 (yellow wire) to node 1 of the circuit. Also, connect scope channel 1 positive (orange wire) to node 1. In addition, connect the scope channel 1 negative (orange-white wire) to node 0. Moreover, connect the ground (black wire) to node 0 of the circuit. The scope channel 1 measures the voltage of the source. Furthermore, connect scope channel 2 positive (blue wire) to node 3 and connect scope channel 2 negative

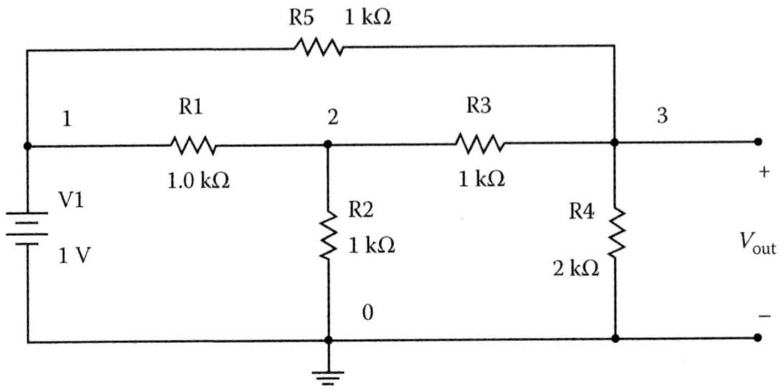

FIGURE 2.8
Bridge-T circuit.

(blue-white wire) to node 0. The scope channel 2 measures the voltage across resistor R4.

Signal Generation:
Click WaveGen in the main WaveForms screen. Go to Basic → dc → Offset (set to 1 V) → Click Run AWG1.

Activation of Voltmeter:
Activate the voltmeter by clicking Voltmeter under the More Instruments tab. Click Enable to start reading the voltages across the source voltage and resistor R4. Figure 2.9 shows the WaveGen screen and the voltmeter readings for input voltage of 1 V. Record your measurements.

Use the voltmeter to measure the input and output voltages. Increase the source voltage from 1 to 5 V with increments of 1 V. Record your measurements. Table 2.5 shows the source voltages and the corresponding voltages at the output (nodes 3 and 0) measured with the voltmeter.

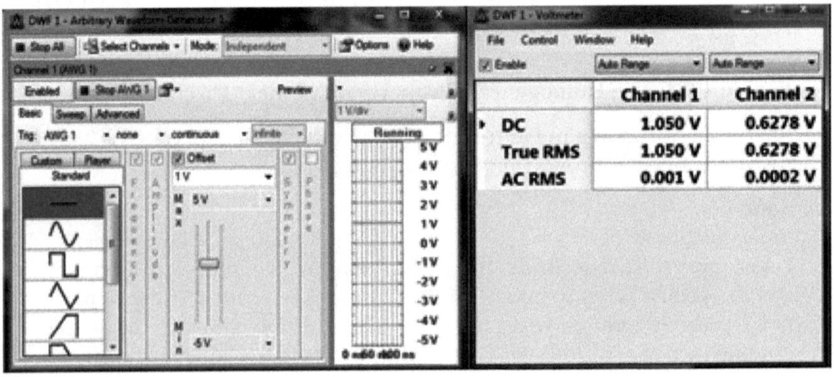

FIGURE 2.9
Power supply and voltmeter readings.

TABLE 2.5
Input and Output Voltages

Source Voltage (V)	Output Voltage (V)	Ratio of Output Voltage versus Input Voltage
1.004	0.624	0.622
2.000	1.243	0.622
2.995	1.863	0.622
3.989	2.482	0.622
4.980	3.100	0.622

It can be seen that ratio of the output voltage with respect to the input voltage is a constant (0.622).

2.5 Thevenin's and Norton's Theorems

Thevenin's Theorem states that "any linear two-terminal circuit containing several voltages and resistances can be replaced by just one single voltage source, V_{Th}, connected in series with a single resistance, R_{Th}." Thevenin's voltage, V_{Th}, is the open-circuit voltage at the terminals, and Thevenin's resistance, R_{Th}, is the equivalent input resistance at the terminals when the independent sources are turned off.

Norton's Theorem states that "any linear two-terminal circuit containing several voltages and resistances can be replaced by just one single current source, I_N, connected in parallel with a single resistance, R_N." Norton's current, I_N, is the short-circuit current at the terminals, and Norton's resistance, R_N, is the equivalent input resistance at the terminals when the independent sources are turned off.

Thevenin's and Norton's Theorems are related through the following expressions

$$R_{Th} = R_N \tag{2.4}$$

$$V_{Th} = I_N * R_N \tag{2.5}$$

Example 2.5 shows how to use the Analog Discovery board to obtain the Thevenin's and Norton's equivalent circuits.

Example 2.5 Thevenin's and Norton's Equivalent Circuit

Determine the Thevenin equivalent circuit at nodes 5 and 0 for the circuit shown in Figure 2.10.

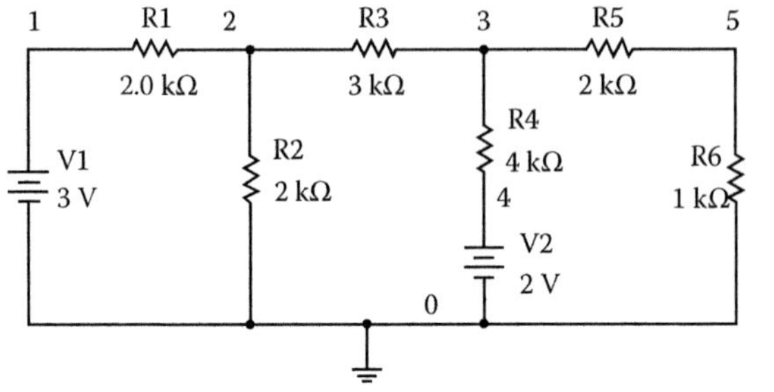

FIGURE 2.10
Circuit with two voltage sources.

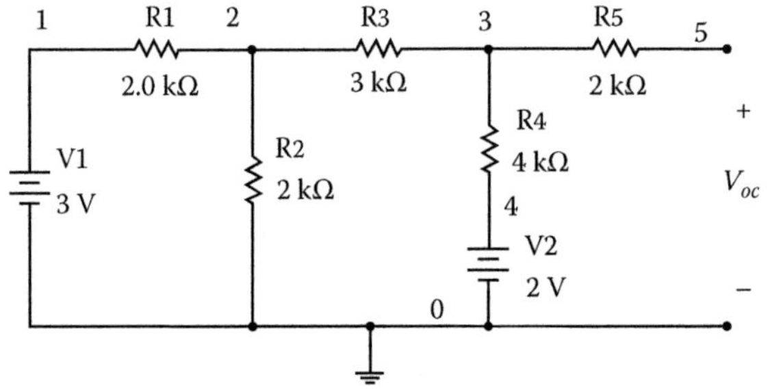

FIGURE 2.11
Circuit for obtaining open-circuit voltage.

Solution:
To obtain the Thevenin's voltage, the circuit is open-circuited at nodes 5 and 0. It is shown in Figure 2.11.

To obtain the open-circuit voltage, the following can be done:

Build the Circuit:
Use a breadboard to build the circuit shown in Figure 2.11. Connect AWG1 (yellow wire) to node 1 of the circuit. Also, connect the ground (black wire) to node 0 of the circuit. In addition, connect AWG2 (yellow-white wire) to node 4 of the circuit.

Signal Generation:
Click WaveGen in the main WaveForms screen. Go to Basic → dc → Offset (set to 3 V) → Click Run AWG1. In addition, in the WaveGen screen, select Channel 2, go to Basic → dc → Offset (set to 2 V) → Click Run AWG2. Figure 2.12 shows the waveforms generated by the arbitrary waveform generators, AWG1 and AWG2.

Measure the voltage between nodes 5 and 0 by using a multimeter. The measured voltage is 1.74 V.

To measure the Thevenin's resistance, short-circuit the voltage sources. The circuit shown in Figure 2.13 is obtained after short-circuiting the voltage sources. The measured resistance value is 3.94 kΩ.

The Thevenin equivalent circuit is shown in Figure 2.14.

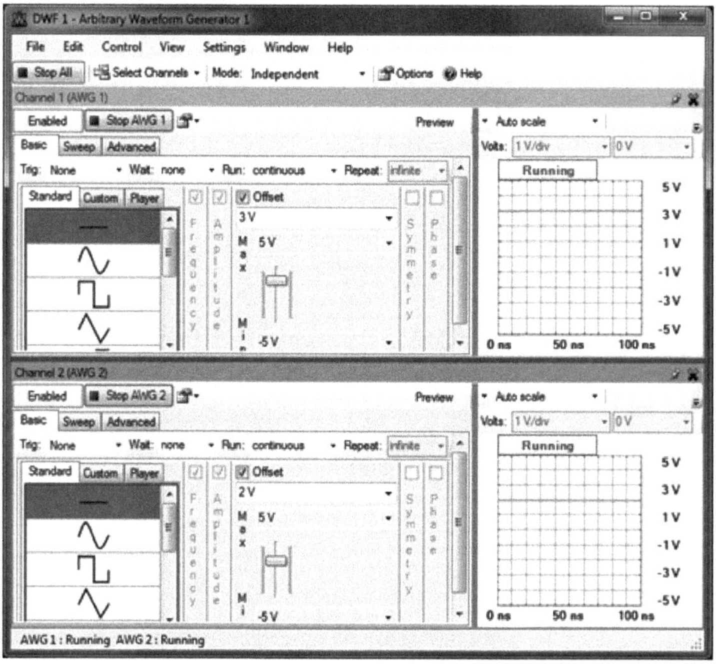

FIGURE 2.12
Signals generated arbitrary waveform generators AWG1 and AWG2.

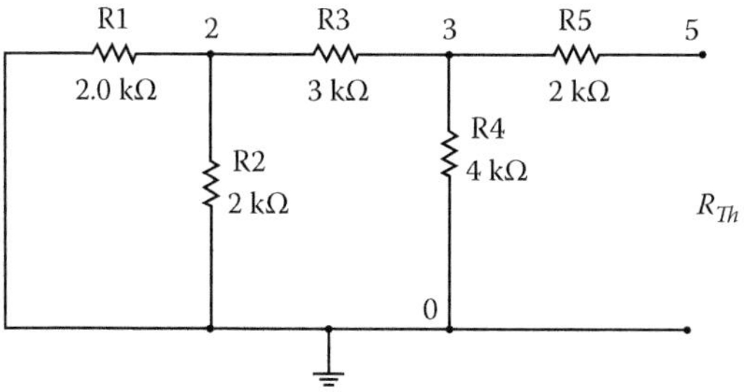

FIGURE 2.13
Circuit of Figure 2.10 without the voltage sources.

PROBLEMS

Problem 2.1 Build the circuit shown in Figure P2.1. (1) Use the Analog Discovery voltmeter to measure the voltages across resistors, R1 and R2. (2) Use the voltage divider rule to calculate the voltages across R1 and R2. (3) Compare the results obtained in parts (1) and (2).

Problem 2.2 Build the circuit shown in Figure P2.2. (1) Use the Analog Discovery board to measure the voltage across the resistor R3. (2) Use nodal

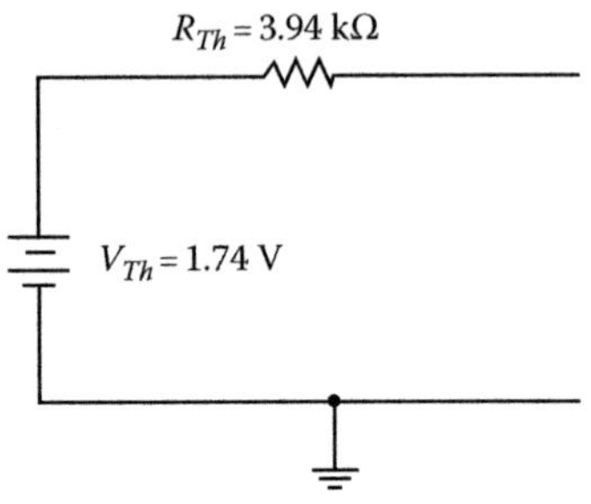

FIGURE 2.14
Thevenin equivalent circuit of Figure 2.10.

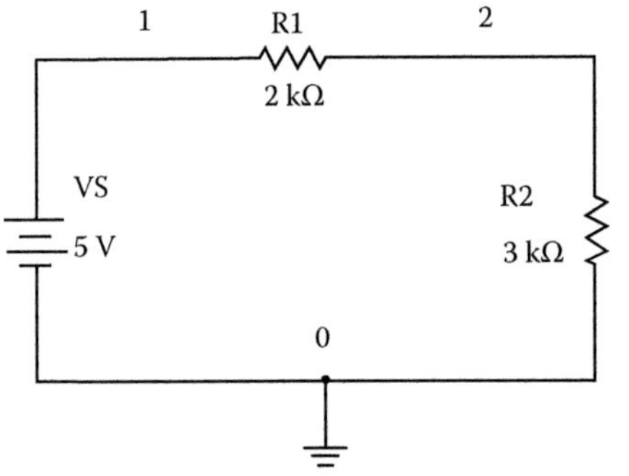

FIGURE P2.1
Circuit with two resistors and a voltage source.

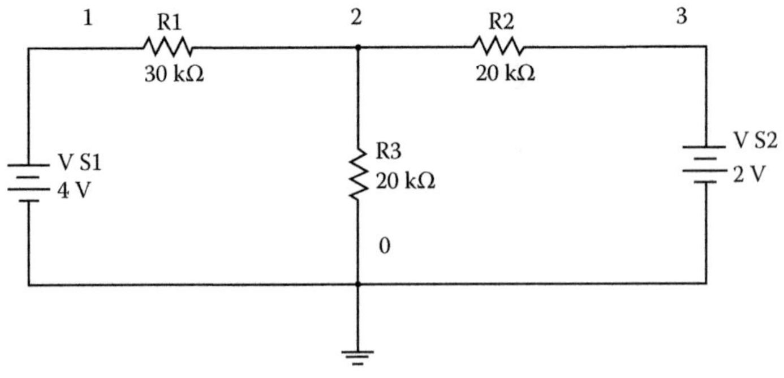

FIGURE P2.2
Circuit with two voltage sources.

analysis to obtain the voltage across R3. (3) Compare the measured and the calculated value of the voltage across R3.

Problem 2.3 Build the circuit of Figure P2.3. (1) Use the Analog Discovery board to measure the voltage across resistors R1 and R2. (2) Based on the

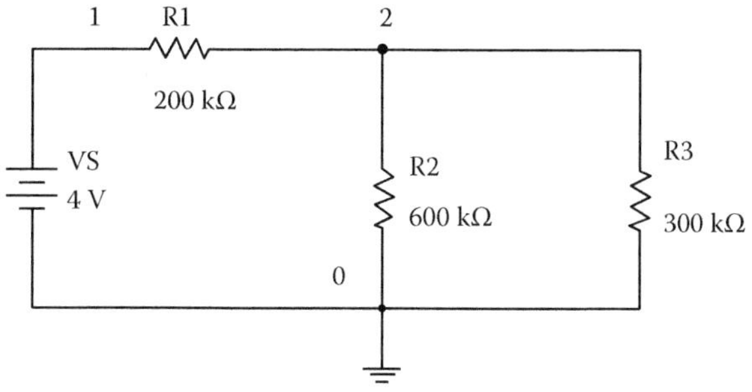

FIGURE P2.3
A resistive circuit.

measured values of the voltages in part (1), use Ohm's Law to calculate the currents flowing through resistors R1, R2, and R3. (3) Use the currents obtained in part (2) to validate the Kirchhoff's Current Law at node 2 of the circuit.

Problem 2.4 Build the circuit shown in Figure P2.4, and obtain the voltage across R6. (2) Decrease the voltage VS from 5 to 1 V with decrement of 1 V, and obtain the voltage across R6. (3) From the results obtained in parts (1) and (2), show that the voltage across the resistor R6 is linearly proportional to the source voltage, VS.

Problem 2.5 (1) Build the circuit shown in Figure P2.5, and use the Analog Discovery board to obtain the voltage across the resistor R6. (2) Set the voltage VS1 = 0 V, VS2 = 2 V, and determine the voltage across R6. (3) Set the voltage VS1 = 3 V, VS2 = 0 V, and determine the voltage across R6. (4) Use the results of parts (1), (2), and (3) to demonstrate the Superposition Theorem.

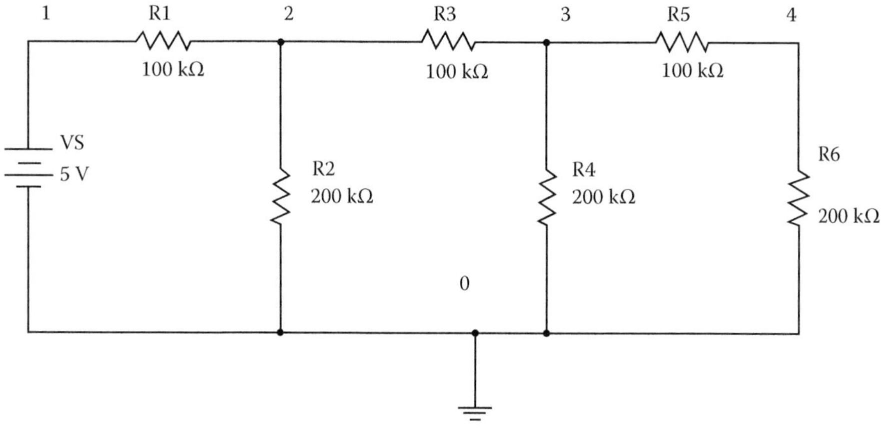

FIGURE P2.4
Ladder network.

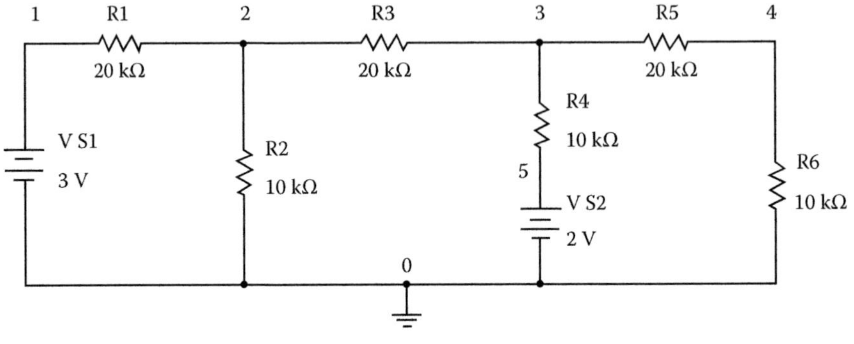

FIGURE P2.5
Circuit for demonstrating superposition theorem.

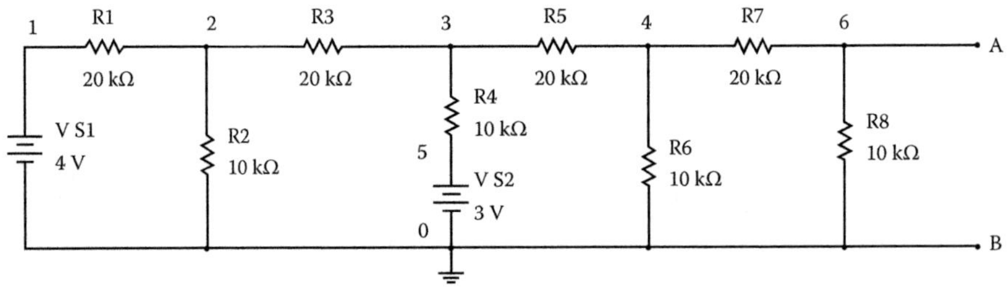

FIGURE P2.6
Circuit with two voltage sources.

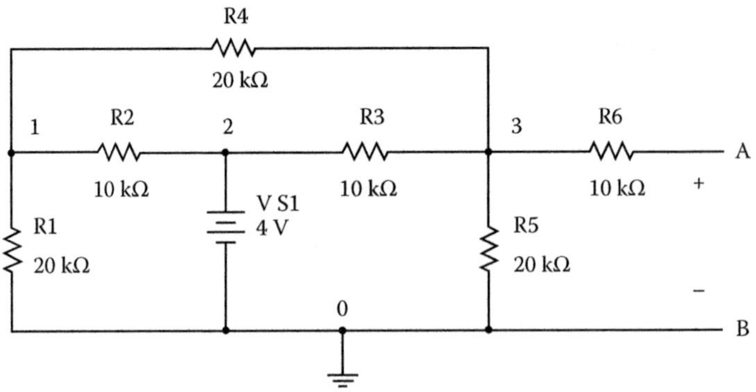

FIGURE P2.7
Bridge-T circuit with a voltage source.

Problem 2.6 For the circuit shown in Figure P2.6, find the Thevenin equivalent circuit between nodes A and B.

Problem 2.7 Find the Norton equivalent circuit of Figure P2.7 at nodes A and B.

3

Transient Analysis

INTRODUCTION In this chapter, the transient responses of RC, RL, and RLC circuits are explored by using the Analog Discovery board. The topics to be explored include (1) charging and discharging capacitors in RC circuits, (2) charging and discharging inductors in RL circuits, and (3) the transient response of RLC circuits.

3.1 RC Circuits

For the RC circuit, shown in Figure 3.1, if the input is a square wave, the capacitor will charge and discharge.

For an RC circuit like the one shown in Figure 3.1, if one uses the KCL and assumes that the initial voltage across the capacitor is zero, one can obtain the differential equation shown in Equation 3.1

$$C\frac{dV_C(t)}{dt} + \frac{V_C - V_S}{R} = 0 \tag{3.1}$$

The expression for charging the capacitor is given as

$$V_C(t) = V_S\left(1 - e^{\frac{-t}{RC}}\right) \tag{3.2}$$

and that for discharging of the capacitor is

$$V_C(t) = V_C(0)e^{\frac{-t}{RC}} \tag{3.3}$$

where

$V_C(0)$=initial voltage across the capacitor

One can quickly charge and discharge a capacitor by using a function generator to generate the square wave. In addition, one can control the amplitude and period *T* of the square wave. The period and the repetition frequency are related by the expression

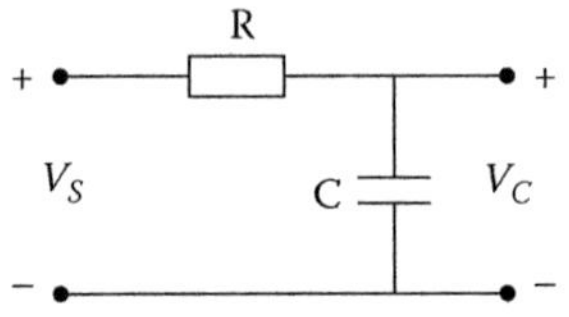

FIGURE 3.1
RC circuit.

$$f = \frac{1}{T} \tag{3.4}$$

where f is the repetition frequency and T is the period of the square wave.

For the RC circuit, the time constant is given as τ=RC. The time constant determines the rate of charge or discharge of a capacitor. After τ seconds, the voltage will have decreased to e^{-1} (about 0.368) of its initial value. After 2τ seconds, it will have decreased to e^{-2} (0.135) its initial value, and after 5τ seconds, the voltage will have decreased to 0.0067 of the initial value. To allow the capacitor to charge to its maximum voltage, it is recommended that the period of the square wave should be far greater than five times the time constant. Example 3.1 shows how to use the Analog Discovery board to explore the behavior of an RC circuit.

Example 3.1 Charging and Discharging of a Capacitor

For the circuit shown in Figure 3.2, the input square wave has a frequency of 1 kHz, the peak-to-peak value of the square wave is 0 to 4 V with an offset voltage of 2 V. Use the Analog Discovery board to obtain the output signal. (1) Measure the time constant of the circuit. (2) Compare the measured time constant with the theoretically calculated value. (3) If the frequency of square wave is increased to 10 kHz without changing the element values, what will be the output voltage?

Solution:

Build the Circuit:

Use a breadboard to build the circuit shown in Figure 3.2. Connect AWG1 (yellow wire) to node 1 of the circuit. Also, connect the scope channel 1

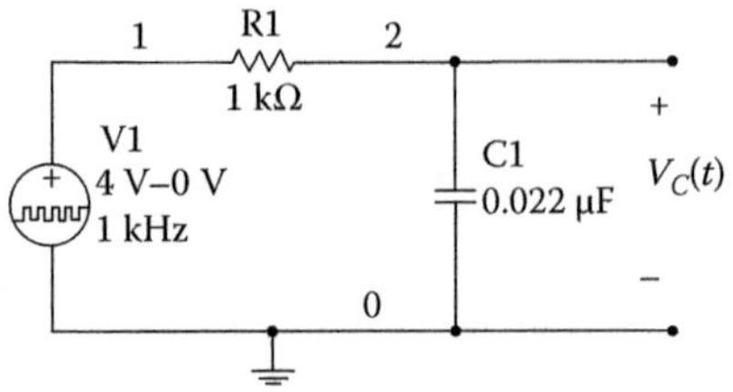

FIGURE 3.2
RC circuit for Example 3.1.

positive (orange wire) to node 1. In addition, connect the scope channel 1 negative (orange-white wire) to node 0. Moreover, connect the ground (black wire) to node 0 of the circuit. Furthermore, connect the scope channel 2 positive (blue wire) to node 2, and the scope channel 2 negative (blue-white wire) to node 0. Scope channel 1 will display the input square wave signal, and scope channel 2 will show the voltage across the capacitor C1.

Generation of Square Wave:
Click on the WaveGen in the main WaveForms screen. Go to Basic → Square → Frequency (set to 1 kHz) → Amplitude (set to 2 V) → Offset (set to 2 V) → Symmetry (set to 50%) → Click on "Run AWG1." Figure 3.3 shows the Analog Discovery WaveGen setup.

Activation of Scope:
Click on the Scope in the main WaveForms screen. Go to Run → Autoset. Make sure the transient waveform of $V_C(t)$ is clearly displayed while the capacitor is charging and discharging.

The input (scope channel 1) and output (scope channel 2) waveforms are shown in Figure 3.4. The time constant is obtained by determining the time it takes for the output voltage to reach 0.637 of the maximum output voltage of 4 V (when the capacitor is charging.)

The measured time constant is 19 μs.

The calculated time constant = RC = 22 μs.

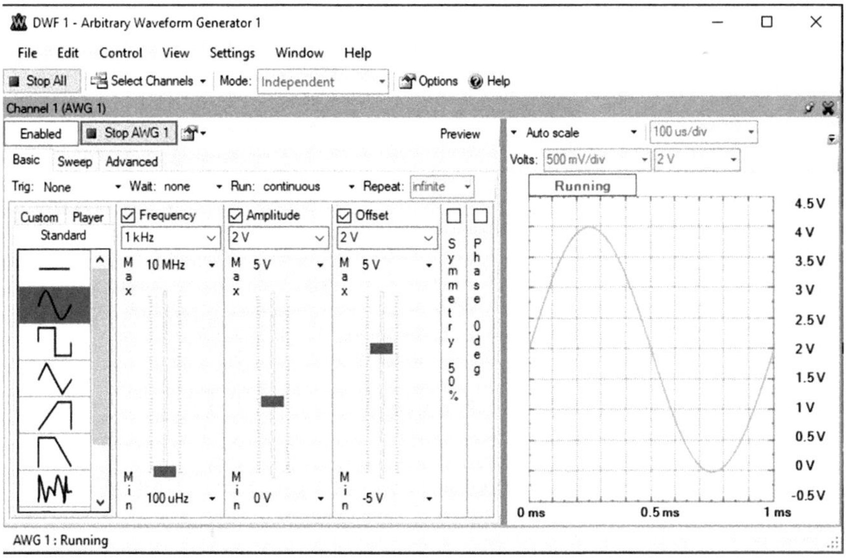

FIGURE 3.3
WaveGen setup for RC circuit.

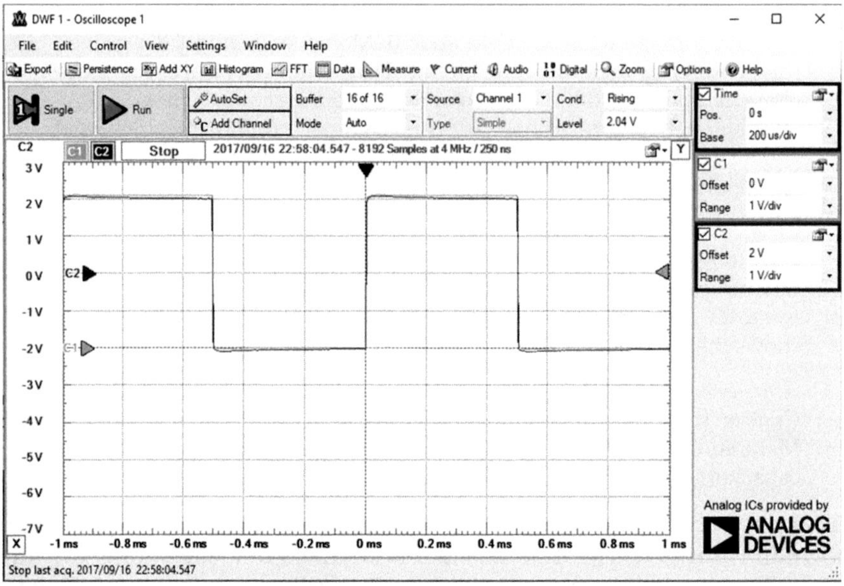

FIGURE 3.4
The input (scope channel 1) and output (scope channel 2) waveforms.

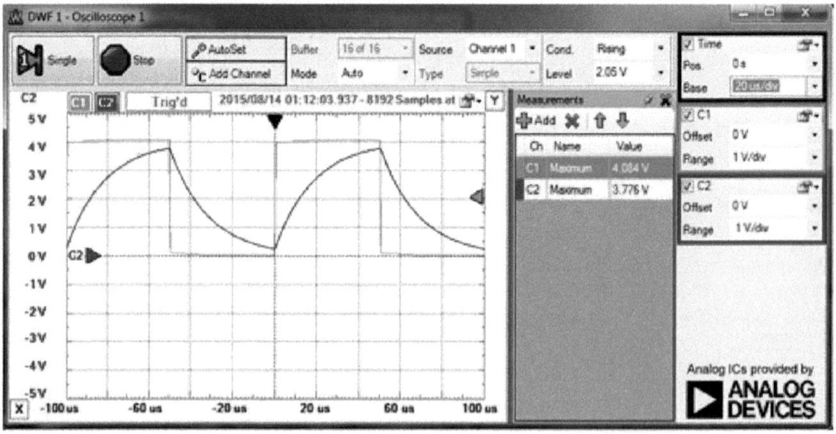

FIGURE 3.5
Input (scope channel 1) and output (scope channel 2) waveforms of RC circuit when square frequency is 10 kHz.

The differences between the calculated and measured time constant might be due to tolerance of the element values.

If the frequency of the square wave generator is increased to 10 kHz, while keeping the $R=1\ \text{k}\Omega$ and $C=0.022\ \mu\text{F}$, input (scope channel 1) and output (scope channel 2) waveforms are shown in Figure 3.5.

3.2 RL Circuit

An RL circuit with source voltage $v(t) = V_S$ is shown in Figure 3.6.

Using KVL, we get

$$L\frac{di(t)}{dt} + Ri(t) = V_S \tag{3.5}$$

If the initial current flowing through the series circuit is zero, the solution of Equation 3.5 is

$$i(t) = \frac{V_S}{R}\left(1 - e^{-\left(\frac{Rt}{L}\right)}\right) \tag{3.6}$$

The voltage across the resistor is

$$\begin{aligned} v_R(t) &= Ri(t) \\ &= V_S\left(1 - e^{-\left(\frac{Rt}{L}\right)}\right) \end{aligned} \tag{3.7}$$

The voltage across the inductor is

$$\begin{aligned} v_L(t) &= V_S - v_R(t) \\ &= V_S e^{-\left(\frac{Rt}{L}\right)} \end{aligned} \tag{3.8}$$

It can be shown that the equation for discharging the inductor is given as

$$i_L(t) = i_L(0)e^{\frac{-Rt}{L}} \tag{3.9}$$

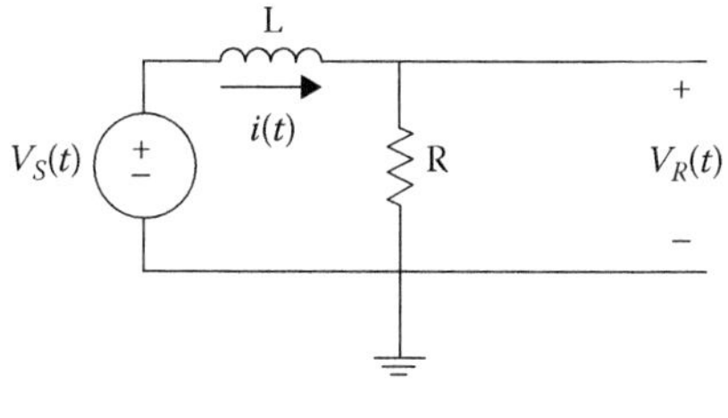

FIGURE 3.6
RL circuit with a voltage source.

where

$i_L(0)$ = initial current flowing through the inductor.

One can quickly charge and discharge an inductor by using a square wave generated by a function generator. The period T and the repetition frequency f of the square wave are related by the expression

$$f = \frac{1}{T} \tag{3.10}$$

For an RL circuit, the time constant is given as $\tau = L/R$. The time constant determines the rate of charging or discharging of a capacitor. After τ seconds, the current will have decreased to e^{-1} (about 0.368) of its initial value. After 2τ seconds, it will have decreased to e^{-2} (0.135) of its initial value, and after 5τ seconds, the current will have decreased to 0.0067 of the initial value. Example 3.2 shows how to use the Analog Discovery board to explore the behavior of an RL circuit.

Example 3.2 Charging and Discharging of an Inductor

For the RL circuit shown in Figure 3.7, the input square wave has a frequency of 10 kHz, the peak-to-peak value of the square wave is 0 to 4 V with offset voltage of 2 V. Use the Analog Discovery board to obtain the output signal. (1) Measure the time constant of the circuit. (2) Compare the measured time constant with the theoretically calculated value. (3) If the frequency of square wave is increased to 100 kHz without changing the element values, what will be the output voltage?

Solution:

Build the Circuit:

Use a breadboard to build the circuit shown in Figure 3.7. Connect the AWG1 (yellow wire) to node 1 of the circuit. Also, connect the scope channel 1 positive (orange wire) to node 1. In addition, connect the scope channel 1 negative (orange-white wire) to node 0. Moreover, connect the ground (black wire) to node 0 of the circuit. Furthermore, connect scope channel 2 positive (blue wire) to node 2, and the scope channel 2 negative (blue-white wire) to node 0. Scope channel 1 will display the

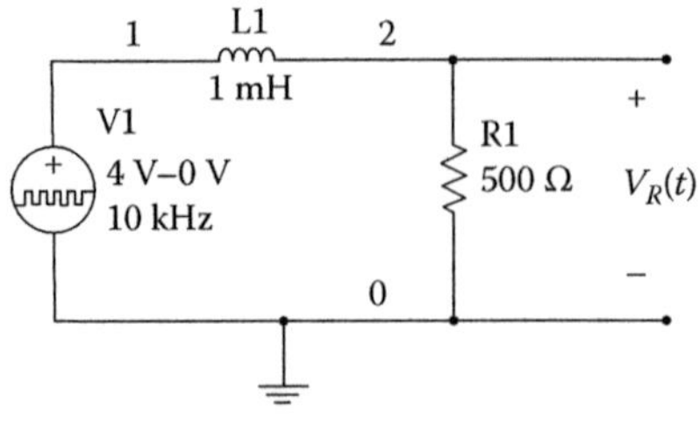

FIGURE 3.7
RL circuit for Example 3.2.

input square wave signal, and scope channel 2 will show the voltage across the resistor R1.

Generation of Square Wave:
Click on the WaveGen in the main WaveForms screen. Go to Basic → Square → Frequency (set to 10 kHz) → Amplitude (set to 2 V) → Offset (set to 2 V) → Symmetry (set to 50%) → Click Run AWG1. Figure 3.8 shows the Analog Discovery WaveGen setup for generating the square wave.

Activation of Scope:
Click on the Scope in the main WaveForms screen. Go to Run → Autoset → Measure → Add → Channel 1 Horizontal → Frequency → Add → Channel 1 Vertical → Maximum → Add → Channel 2 Vertical → Maximum → Add → Channel 2 Vertical → Minimum → Add Selected Measurement → Close Add Instrument.

The input (scope channel 1) and output (scope channel 2) waveforms are shown in Figure 3.9. The time constant is obtained by determining the time it takes for the output voltage to reach 0.637 of the maximum output voltage of 4 V.

The measured time constant is 1.9 μs.

The calculated time constant $= L/R = 2.0\,\mu s$.

The differences between the calculated and measured time constant might be due to tolerance of the element values.

If the frequency of the square wave generator is increased to 100 kHz, while keeping $R = 500\ k\Omega$ and $L = 1.0\ mH$, input (scope channel 1) and output (scope channel 2) waveforms are shown in Figure 3.10.

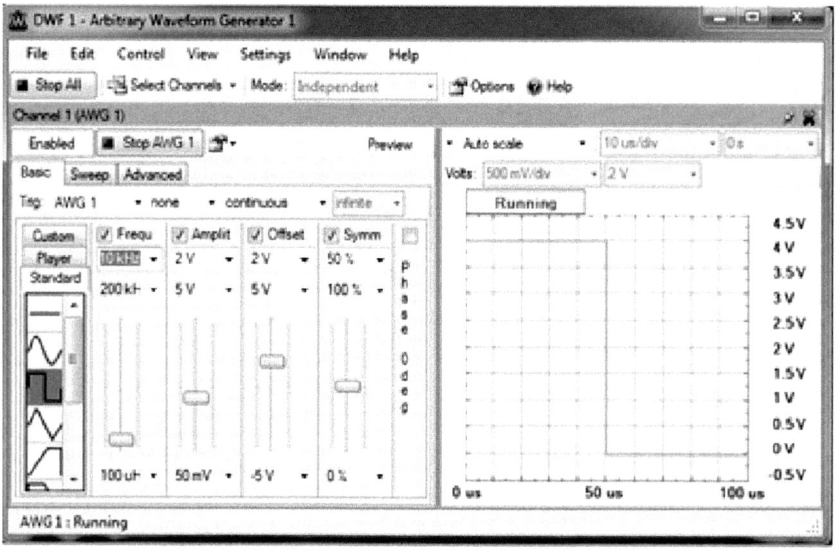

FIGURE 3.8
WaveGen setup generating a square wave.

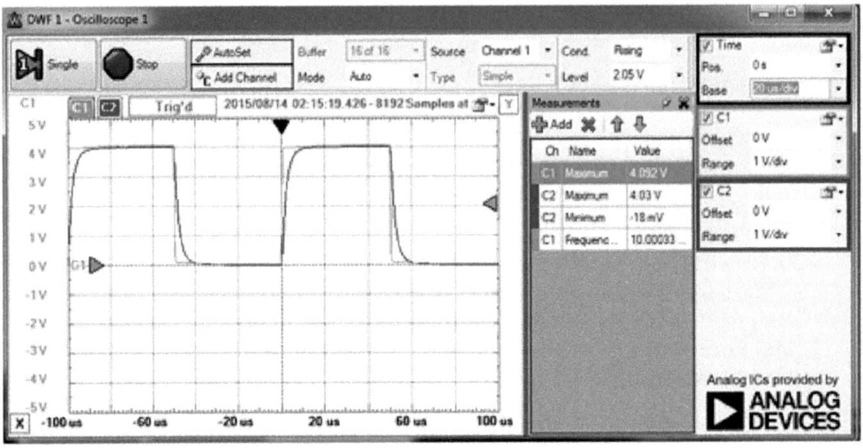

FIGURE 3.9
The input and output waveforms of the RL circuit.

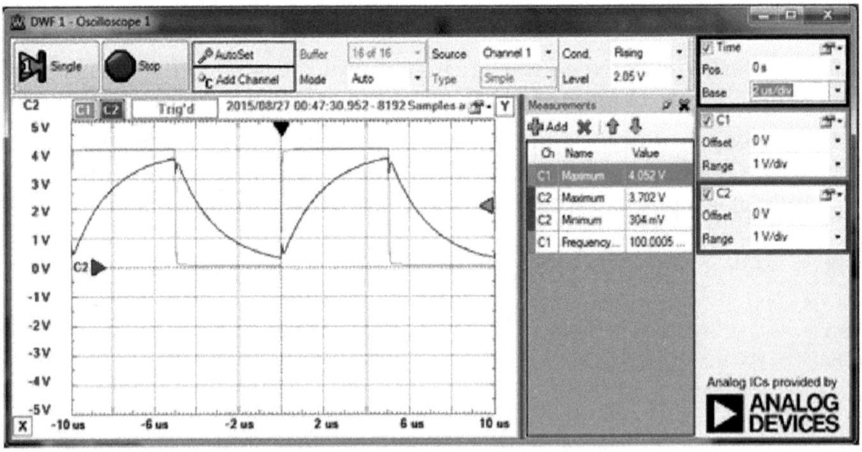

FIGURE 3.10
The input (scope channel 1) and output (scope channel 2) waveforms of the RL circuit for frequency of 100 kHz.

3.3 RLC Circuits

For the series RLC circuit shown in Figure 3.11, one can use KVL to obtain Equation 3.11.

$$v_S(t) = L\frac{di(t)}{dt} + \frac{1}{C}\int_{-\infty}^{t} i(\tau)\, d\tau + Ri(t) \tag{3.11}$$

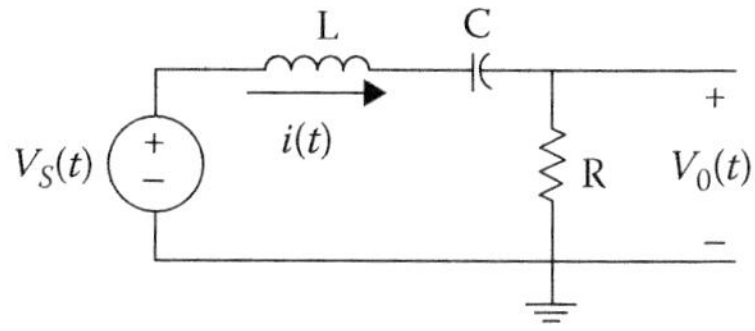

FIGURE 3.11
Series RLC circuit.

Differentiating the above expression, we get

$$\frac{dv_S(t)}{dt} = L\frac{d^2i(t)}{dt^2} + R\frac{di(t)}{dt} + \frac{i(t)}{C}$$

i.e.,

$$\frac{1}{L}\frac{dv_S(t)}{dt} = \frac{d^2i(t)}{dt^2} + \frac{R}{L}\frac{di(t)}{dt} + \frac{i(t)}{LC} \tag{3.12}$$

The homogeneous solution can be found by making $v_S(t)$=constant, thus

$$0 = \frac{d^2i(t)}{dt^2} + \frac{R}{L}\frac{di(t)}{dt} + \frac{i(t)}{LC} \tag{3.13}$$

The characteristic equation is

$$0 = S^2 + 2\alpha S + w_0^2 \tag{3.14}$$

where

$$\alpha = \frac{R}{2L} \text{ and} \tag{3.15}$$

α is damping factor

$$w_0 = \frac{1}{\sqrt{LC}} \tag{3.16}$$

w_0 is undamped frequency

The roots of the characteristic equation, S_1 and S_2, can be determined. If we assume that the roots are

$$S_1 = -\alpha + \sqrt{\alpha^2 - \omega_0^2} \tag{3.17}$$

$$S_2 = -\alpha - \sqrt{\alpha^2 - \omega_0^2} \tag{3.18}$$

the solution to the homogeneous solution is

$$i_h(t) = A_1 e^{S_1 t} + A_2 e^{S_2 t} \tag{3.19}$$

where A_1 and A_2 are constants.

If $v_S(t)$ is a constant, then the forced solution will also be a constant and be given as

$$i_f(t) = A_3 \tag{3.20}$$

The total solution is given as

$$i(t) = A_1 e^{S_1 t} + A_2 e^{S_2 t} + A_3 \tag{3.21}$$

where A_1, A_2, and A_3 are obtained from initial conditions.

The roots of Equation 3.21 are given by Equations 3.17 and 3.18.

Three response can result from Equation 3.21. They are as follows:

Case 1: Overdamped response occurs when $\alpha \succ \omega_0$. The roots are real and the expression for $i(t)$ is given as

$$i(t) = B_1 e^{S_1 t} + B_2 e^{S_2 t} + B_3 \tag{3.22}$$

where B_1, B_2, B_3 are obtained from initial conditions.

Case 2: Critically damped response occurs when $\alpha = \omega_0$. There are two real roots that are equal, and the expression for $i(t)$ is given as

$$i(t) = \left(C_1 + C_2 t\right) e^{-\alpha t} + C_3 \tag{3.23}$$

where C_1, C_2, C_3 are obtained from initial conditions.

Case 3: Underdamped response occurs when $\alpha \prec \omega_0$. The roots are complex and the expression for $i(t)$ is given as

$$i(t) = \left(D_1 \cos(w_d t) + D_2 \sin(w_d t)\right) e^{-\alpha t} + D_3 \tag{3.24}$$

where

$$\omega_d = \sqrt{\omega_0^2 - \alpha^2} \tag{3.25}$$

ω_d is damped frequency

and D_1, D_2, D_3 are obtained from initial conditions.

In the laboratory, two points are obtained on the envelope of the damped exponential sinusoid to calculate the damping factor and damping frequency. If the first positive local maximum of the sinusoid is at the point (t_1, V_1) and the second positive local maximum of the sinusoid is at the point (t_2, V_2), where t_1, t_2 are measurements of time corresponding to envelope voltages V_1, V_2. The damping factor can be calculated as

$$\alpha = \frac{\ln(V_1) - \ln(V_2)}{t_2 - t_1} \tag{3.26}$$

Also, if the period of the damped exponential sinusoid is given as

$$T_d = t_2 - t_1 \tag{3.27}$$

The damped frequency is given as

$$\omega_d = \frac{2\pi}{T_d} \tag{3.28}$$

Due to the duality principle for electric circuits, the equations for the parallel RLC are similar to the series RLC circuit. However, in the case of the parallel RLC ciruit, the damping factor is given as

$$\alpha = \frac{1}{2RC} \tag{3.29}$$

$$w_0 = \frac{1}{\sqrt{LC}} \tag{3.30}$$

Example 3.3 illustrates the use of the Analog Discovery board to obtain the transient response of a RLC circuit.

Example 3.3 Transient Response of RLC Circuit

For the RLC circuit shown in Figure 3.12, the input square wave has a frequency of 200 Hz, the peak-to-peak value of the square wave is 0 to 4 V with offset voltage of 2 V. (1) Use the Analog Discovery board to obtain the voltage across the resistor. (2) Determine the damping factor and damped frequency. (3) Compare the measured damping factor and damped frequency with the theoretically-calculated values, and (4) when $R = 2000\,\Omega$, $L = 40$ mH, and $C = 0.1\,\mu F$, determine the type of transient response of the output waveform.

***Solution*:**

Build the Circuit:

Use a breadboard to build the circuit shown in Figure 3.12 with R1 = 200 Ω, L1 = 40.0 mH, and C1 = 0.1 μF. Connect the AWG1 (yellow wire) to node 1 of the circuit. Also, connect the scope channel 1 positive (orange wire) to

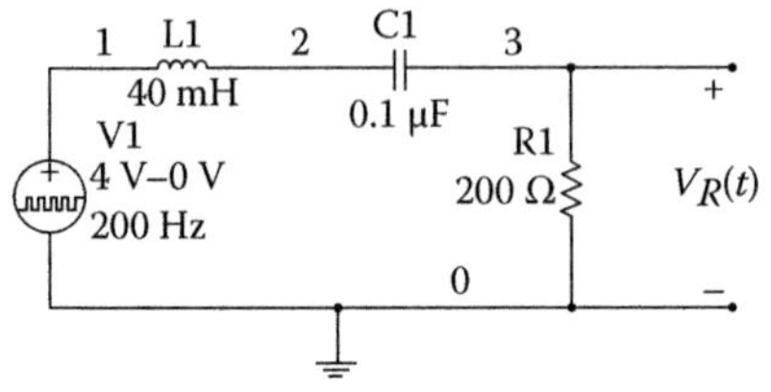

FIGURE 3.12
RLC circuit of Example 3.3.

node 1. In addition, connect the scope channel 1 negative (orange-white wire) to node 0. Moreover, connect the ground (black wire) to node 0 of the circuit. Furthermore, connect the scope channel 2 positive (blue wire) to node 3, and the scope channel 2 negative (blue-white wire) to node 0.

Generation of Square Wave:
Click on the WaveGen in the main WaveForms screen. Go to Basic → Square → Frequency (set to 200 Hz) → Amplitude (set to 2 V) → Offset (set to 2 V) → Symmetry (set to 50%) → Click Run AWG1. The square wave generated by AWG1 is shown in Figure 3.13.

Activation of Scope:
Click on the Scope in the main WaveForms screen. Go to Run → Autoset → Measure → Add → Channel 1 Vertical → Frequency → Add Selected Measurement → Close Add Instrument.

Make sure the transient waveform of $V_R(t)$ is clearly displayed. *Hint*: Click on the *AutoSet* button to automatically rescale the scope axes. Figure 3.13 shows the Digilent Analog Discovery WaveGen setup.

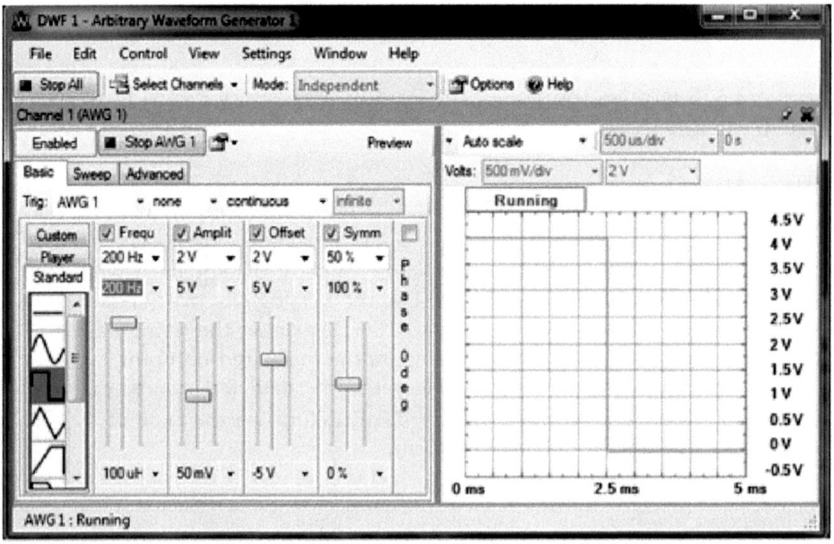

FIGURE 3.13
WaveGen setup for generating a square wave.

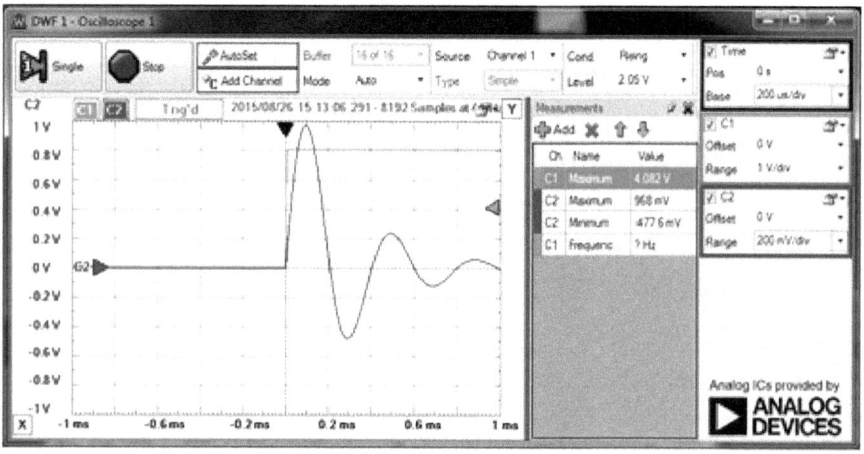

FIGURE 3.14
The input (scope channel 1) and output (scope channel 2) waveforms for the underdamped response.

The input and output waveforms are shown in Figure 3.14. Channel 1 (C1) displays the input square wave signal, and scope channel 2 (C2) shows an underdamped response.

Measure time t_1 and the voltage V_1 of the first positive local maximum of the decaying sinusoid. In addition, measure the time t_2 and the voltage V_2 of the second positive local maximum of the decaying sinusoid. Use Equations 3.26 and 3.28 to calculate the measured damping factor and the damped frequency, respectively.

From Figure 3.14, one can get the following coordinates for the first and second local maximum of the damped sinusoid:

$$(t_1, V_1) = (86.0\ \mu\text{s},\ 0.9794\ \text{V})$$

$$(t_2, V_2) = (48.6\ \mu\text{s},\ 0.244\ \text{V})$$

Using Equations 3.26 and 3.28, the damping factor and damped frequency were calculated. Table 3.1 shows the calculated values and the measured

TABLE 3.1
Calculated and Measured Values of the Underdamped Response of Figure 3.12

Item	Calculated Value	Measured Value
Undamped frequency (rad/s)	1.58×10^4	1.61×10^4
Damped frequency (rad/s)	1.56×10^4	1.57×10^4
Damped frequency (Hz)	2.86×10^3	2.50×10^3
Period of damped frequency (s)	4.02×10^{-4}	4.00×10^{-4}
Damping factor	2500	3474

values of the undamped frequency, damped frequency, damping factor, and period of the damped frequency.

When the 200 Ω resistor was replaced by 2.0 kΩ, while keeping the inductor at 40.0 mH, a capacitor value of 0.1 μF, and 200 Hz square wave, the output waveform obtained is shown in Figure 3.15.

When the frequency of the square wave was increased from 200 to 2000 Hz, the waveforms obtained are shown in Figure 3.16.

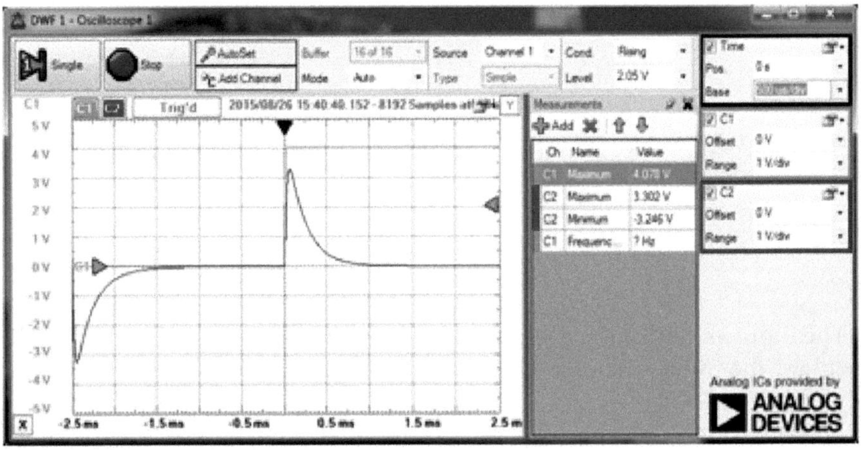

FIGURE 3.15
The input (scope channel 1) and output (scope channel 2) waveforms for the overdamped response.

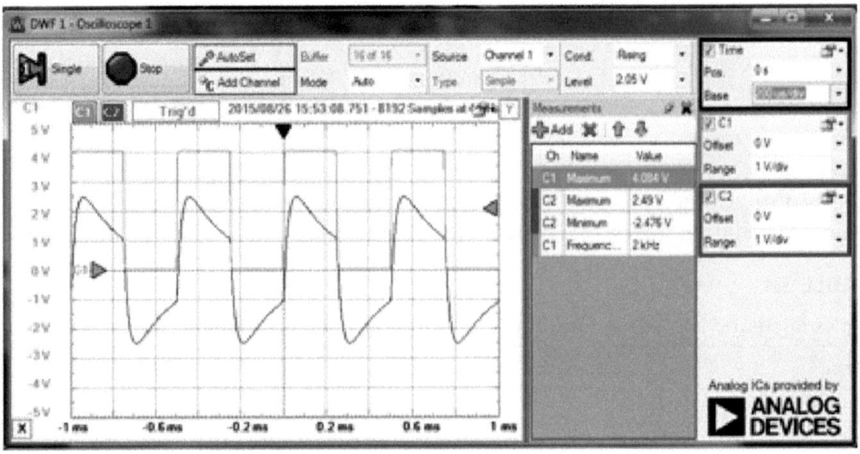

FIGURE 3.16
The input (scope channel 1) and output (scope channel 2) waveforms for the 2000 Hz square wave.

PROBLEMS

Problem 3.1 For the circuit shown in Figure P3.1, the input square wave has a frequency of 100 Hz, and the peak-to-peak value of the square wave is 0 to 4 V with offset voltage of 2 V. (1) Use the Analog Discovery board to obtain the output waveform. (2) Measure the time constant of the circuit. (3) Compare the measured time constant with the theoretically calculated value.

Problem 3.2 If Figure 3.2 has component values of R1 = 1 kΩ and C1 = 1.0 μF, (1) estimate the period (and hence frequency) of the input square wave that will make the output waveform almost a square-wave. (2) Give reasons for your estimated frequency. (3) Build the circuit and use the Analog Discovery board to confirm that the output voltage is almost a square wave when the input frequency of the square wave is equal to your estimated frequency.

Problem 3.3 For the circuit shown in Figure P3.3, the input square wave has a frequency of 200 Hz, and the peak-to-peak value of the square wave is 0 to 4 V with offset voltage of 2 V. (1) Use the Analog Discovery board to obtain the output signal. (2) Measure the time constant of the circuit. (3) Compare the measured time constant with the theoretically calculated value.

Problem 3.4 If Figure 3.6 has component values of $R = 5000\ \Omega$ and $L = 40$ mH, (1) estimate the period (and hence frequency) of the input square wave that will make the output waveform almost a square-wave. (2) Give reasons for your estimated frequency. (3) Build the circuit and use the Analog Discovery

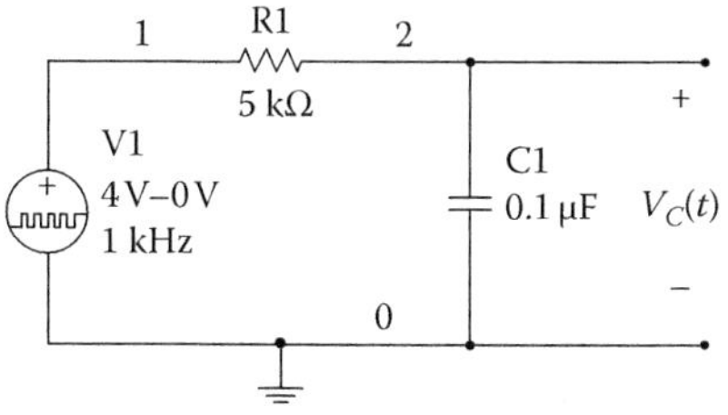

FIGURE P3.1
RC circuit for Exercise 3.1.

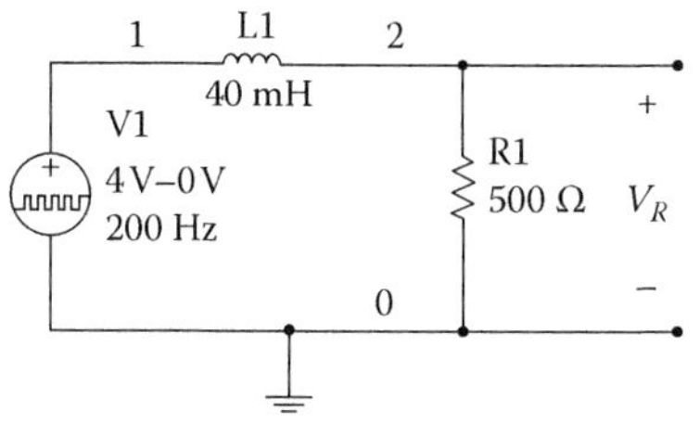

FIGURE P3.3
RL circuit for Exercise 3.3.

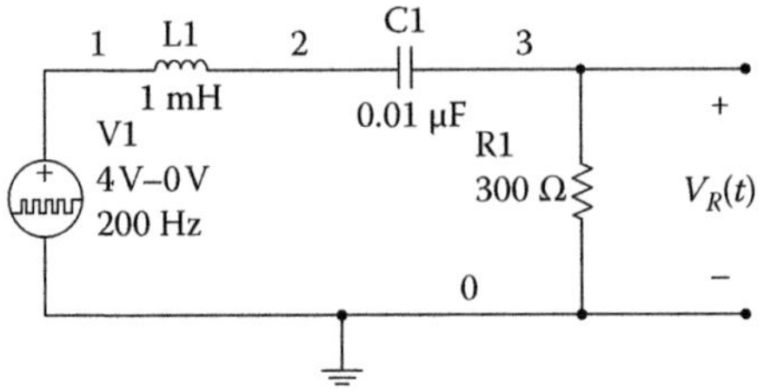

FIGURE P3.5
RLC circuit for Exercise 3.5.

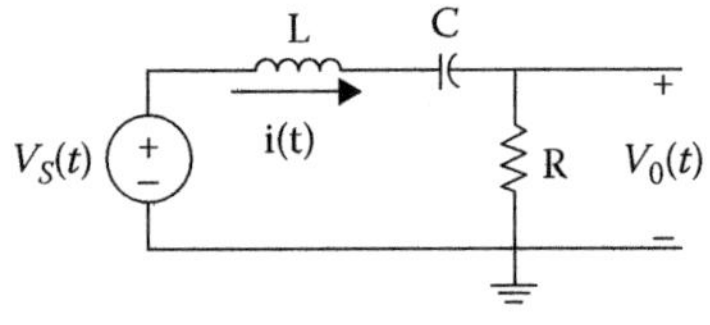

FIGURE P3.6
RLC circuit for Exercise 3.6.

board to confirm that the output voltage is almost a square wave when the input frequency of the square wave is equal to your estimated frequency.

Problem 3.5 For the RLC circuit shown in Figure P3.5, the input square wave has a frequency of 200 Hz and the peak-to-peak value of the square wave is 0 to 4 V with offset voltage of 2 V. Use the Analog Discovery board to obtain the voltage across the resistor. (1) Determine the damping factor and damped frequency. (2) Compare the measured damping factor and damped frequency with the theoretically-calculated values. (3) What might be the causes of errors between the calculated and the measured values?

Problem 3.6 For the series RLC circuit shown in Figure P3.6, $R = 5000\,\Omega$, $L = 40.0\,\text{mH}$, and $C = 0.01\,\mu\text{F}$. (1) Calculate the damping factor and the undamped frequency. (2) Indicate the type of response of the circuit (underdamped, critically damped and overdamped). (3) Build the circuit and use the Analog Discovery board to the output voltage $V_o(t)$. You can use an input square wave of peak-to-peak value of 4 V with dc offset of 2 V and a frequency of your choice.

Problem 3.7 For a series RLC circuit, such as the one shown in Figure P3.6, if $L = 4.0\,\text{mH}$, and $C = 0.001\,\mu\text{F}$, (1) calculate the value of resistor R that will result in critically damped response. (2) Build the circuit and use a resistor that will give you an underdamped response. You can use an input square wave of peak-to-peak value of 4 V with dc offset of 2 V and a frequency of your choice. (3) Measure the damping factor and damped frequency. (4) Compare the measured damping factor and damped frequency with the theoretically-calculated values.

4

Impedance, Power Calculation, and Frequency Response

INTRODUCTION This chapter discusses the determination of impedance of circuit, measurements involving complex power, and the way to obtain the frequency response of networks. While determining impedances, the phase difference between two sinusoids will be discussed. In addition, the rms value of signals will be measured with the Analog Discovery board. Furthermore, the Network Analyzer will be used to obtain the magnitude response of a network.

4.1 Impedance Measurements

Figure 4.1 shows a voltage source $v(t)$ connected to network with impedance Z. The voltage $v(t)$ is a sinusoid given as

$$v(t) = V_m \cos(wt + \theta_V) \tag{4.1}$$

If we assume a linear circuit, the current $i(t)$ will have the same frequency as the voltage, and it will be given as

$$i(t) = I_m \cos(wt + \theta_I) \tag{4.2}$$

The phasor voltage and current are given as

$$V = V_m e^{j\theta_V} = V_m \angle \theta_V \tag{4.3}$$

$$I = I_m e^{j\theta_I} = I_m \angle \theta_I \tag{4.4}$$

The impedance is given as the ratio between the phasor voltage and the phasor current. It is given as

$$Z = \frac{V}{I} = \frac{V_m}{I_m} \angle(\theta_V - \theta_I) \tag{4.5}$$

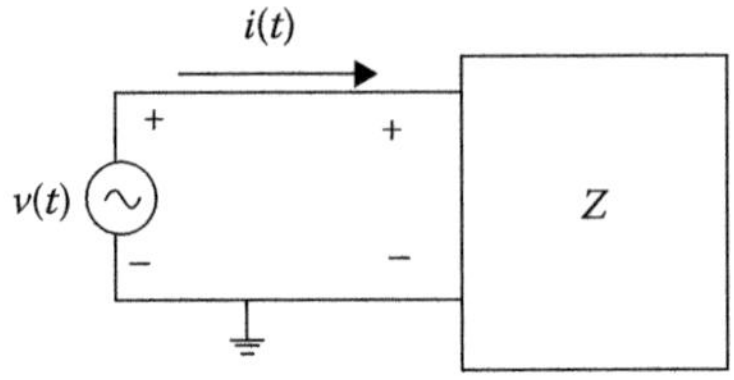

FIGURE 4.1
One-port network with impedance Z.

where

$$Z = Z_m \angle \theta_Z \quad (4.6)$$

$$Z_m = \frac{V_m}{I_m} \quad (4.7)$$

$$\theta_Z = (\theta_V - \theta_I) \quad (4.8)$$

The magnitudes V_m and I_m can be obtained from a scope, and the phase difference $\theta_Z = (\theta_V - \theta_I)$ can also be determined by measuring the phase difference between the voltage and current waveforms.

If t_d is the phase difference between the voltage and current waveforms, and T_P is the period of the voltage or current waveform, then the phase difference is given as

$$\theta_Z = (\theta_V - \theta_I) = \frac{t_d}{T_P}(360°) \quad (4.9)$$

The following steps can be used to obtain the phase difference between two waveforms:

1. Display the two signals on scope channels 1 and 2 as a function of time (see Figure 4.2).
2. Make the offset of channels 1 and 2 zero volt.
3. Use the zero crossings to measure the period, T_P.
4. Measure the smallest time difference between the zero crossings of the two signals, t_d.
5. The phase difference is calculated by using Equation 4.10

$$|\theta_V - \theta_I| = \frac{t_d}{T_P}(360°) \quad (4.10)$$

6. If the voltage signal $v(t)$ attains its maximum value earlier than the current signal $i(t)$, then the phase angle is positive; otherwise, it is negative.

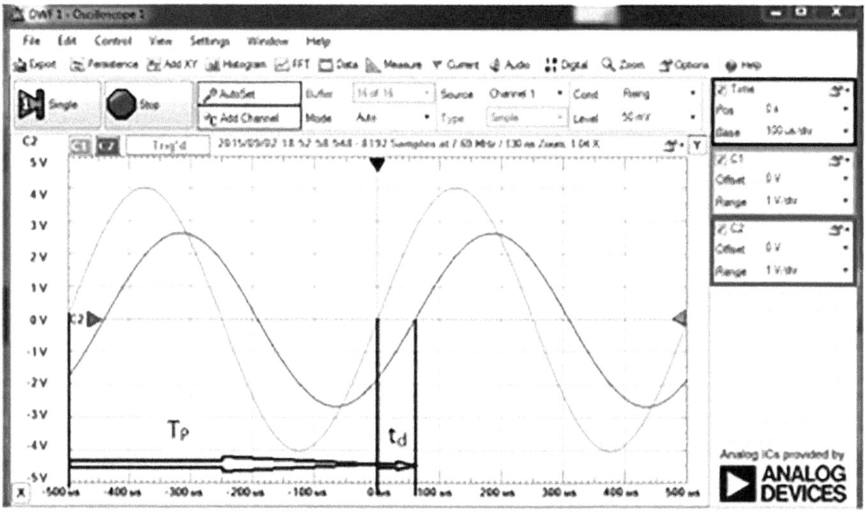

FIGURE 4.2
Display of two signals.

To measure the current, a known value resistor will be placed in series with the circuit under test. Channel 1 of the scope will be connected across the known resistor, and channel 2 will be connected across the circuit under test. The Mathematic Channel can be used to monitor the current by dividing the voltage across the known valued resistor by the resistance of the resistor. Example 4.1 describes how to use the Analog Discovery board to obtain the impedance of a circuit.

Example 4.1 Determining the Impedance of a Circuit

For the circuit shown in Figure 4.3, find the impedance of the circuit from nodes 1 and 0 at the frequency of 5 kHz.

Solution:
A resistor of 100 Ω has been inserted in series with the circuit. The augmented circuit is shown in Figure 4.4.

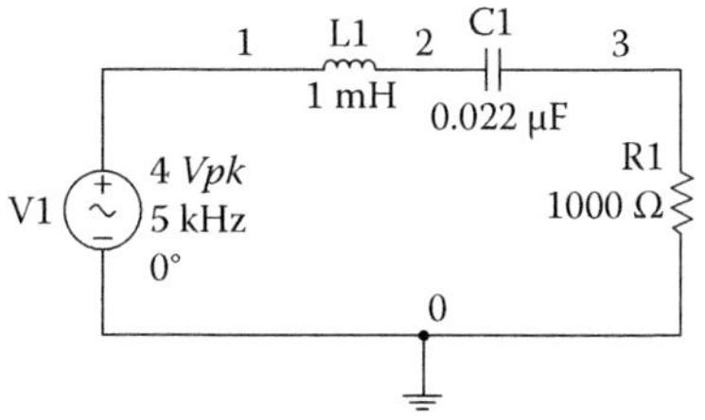

FIGURE 4.3
Impedance of RLC circuit.

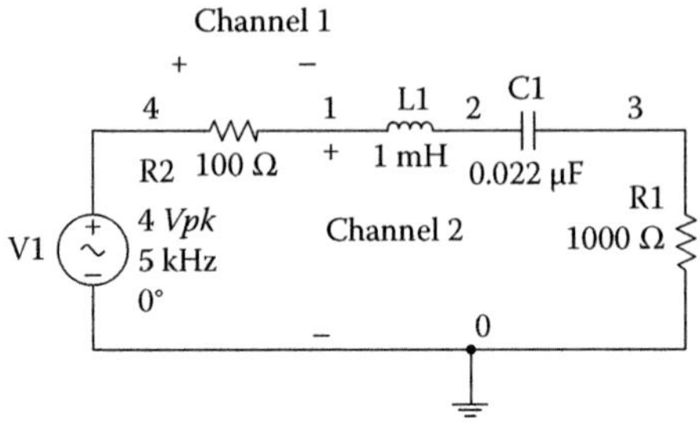

FIGURE 4.4
Augmented version of Figure 4.3 showing 100 Ω resistor to monitor current.

Build the Circuit:
Use a breadboard to build the circuit shown in Figure 4.4. Connect the AWG1 (yellow wire) to node 4 of the circuit. Also, connect the scope channel 1 positive (orange wire) to node 4. In addition, connect the scope channel 1 negative (orange-white wire) to node 1. Moreover, connect the ground (black wire) to node 0 of the circuit. Furthermore, connect the scope channel 2 positive (blue wire) to node 1, and the scope channel 2 negative (blue-white wire) to node 0. Channel 1 measures the voltage across the 100 Ω resistor and channel 2 measures the voltage at the input of the series RLC circuit.

Generation of Sine Wave:
Click on the WaveGen in the main WaveForms screen. Go to Basic → Sinusoid → Frequency (set to 5 kHz) → Amplitude (set to 4 V) → Offset (set to 0 V) → Symmetry (set to 50%) → Click Run AWG1. Figure 4.5 shows the Analog Discovery WaveGen setup.

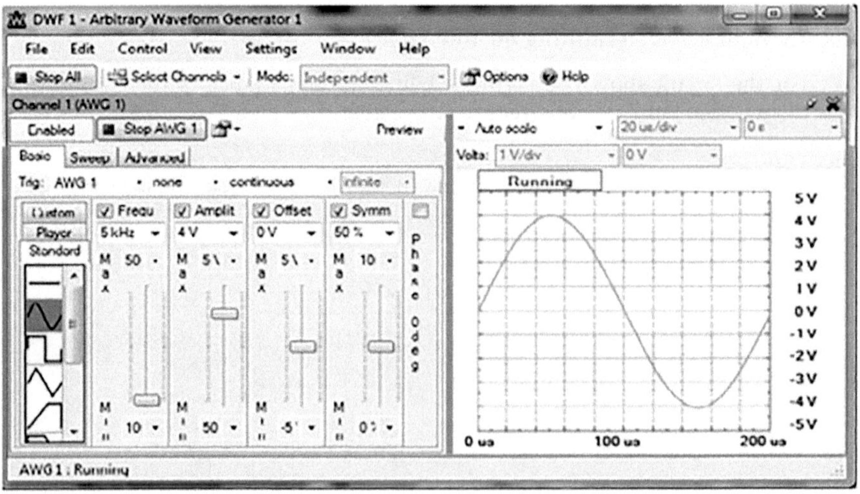

FIGURE 4.5
Analog Discovery WaveGen setup for impedance measurement.

Activation of Scope:
Click on the Scope in the main WaveForms screen. Go to Run → Autoset → Measure → Add → Channel 1 Horizontal → Period → Add → Channel 1 Vertical → peak-to-peak voltage → Add → Channel 2 Vertical → peak-to-peak voltage → Control → Add Mathematic Channel → Custom → C1/100 → Add → Channel M1 Vertical → peak-to-peak voltage → Change Math Mode Units to A → Add Selected Measurement → Close Add Instrument.

From Figure 4.6, we get the following measurements:

Peak-to-peak voltage between nodes 1 and 0 of Figure 4.4 is 7.84 V.
Peak value of the voltage between nodes 1 and 0 of Figure 4.4 is 3.92 V.
Peak-to-peak voltage across 100 Ω resistor of Figure 4.4 is 399.4 mV.
Peak-to-peak current flowing through 100 Ω resistor of Figure 4.4 is 3.994 mA.
Peak value of current flowing through 100 Ω resistor of Figure 4.4 is 1.997 mA.

The magnitude of the impedance is given by Equation 4.7 as

$$Z_m = \frac{V_m}{I_m} = \frac{3.920\ \text{V}}{1.997\ \text{mA}} = 1.963\ \text{K}\Omega \tag{4.11}$$

Measurements made on the scope display show that the time difference between the voltage and current is 34 μs. Using Equation 4.10, we have

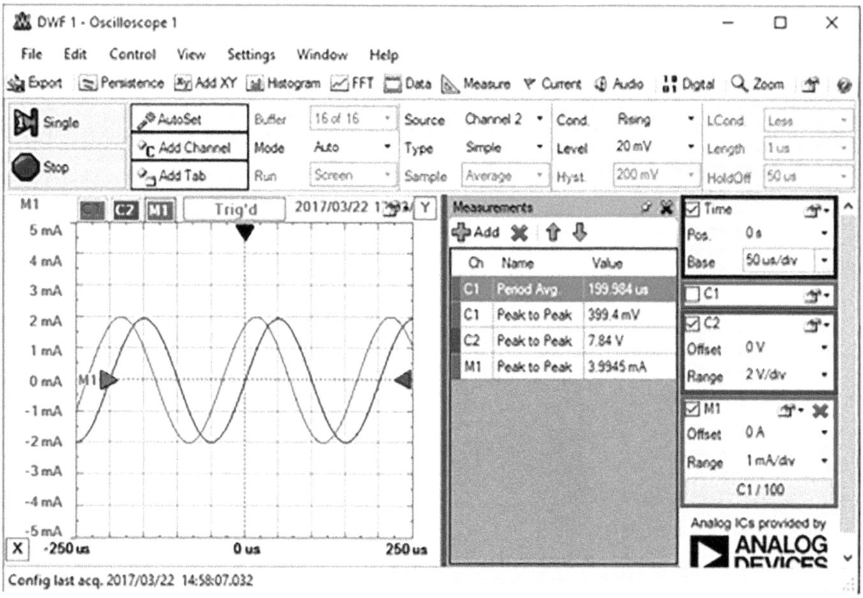

FIGURE 4.6
Scope display of input of RLC circuit (C2) and current flowing through resistor R2.

$$|\theta_V - \theta_I| = \frac{t_d}{T_P}(360°) = \frac{34\ \mu s}{200\ us} \times 360° = 61.2° \tag{4.12}$$

From Figure 4.5, the current waveform (M1) attains its maximum value earlier than that of the voltage waveform (C2). Therefore, the current leads the voltage, and thus

$$\theta_Z = (\theta_V - \theta_I) = -61.2° \tag{4.13}$$

Thus, the measured impedance is given as

$$Z = Z_m \angle \theta_Z = 1963 \angle -61.2°\ \Omega \tag{4.14}$$

For $L = 1.0\ \text{mH}$, $C = 0.022\ \mu\text{F}$, and $\omega = (2\pi)(5000)$ Hz, the calculated impedance is

$$Z_{calculated} = 1000 + j\omega L - \frac{j}{\omega C}$$

$$Z_{calculated} = 1758.9 \angle -55.35° \tag{4.15}$$

It can be seen from Equations 4.14 and 4.15 that the measured and calculated values of the impedance are close to one another.

4.2 Rms Values

If $v(t)$ and $i(t)$ are periodic with period T, the rms, or effective values of the voltage and current, are

$$V_{rms} = \sqrt{\frac{1}{T}\int_0^T v^2(t)\,dt} \tag{4.16}$$

$$I_{rms} = \sqrt{\frac{1}{T}\int_0^T i^2(t)\,dt} \tag{4.17}$$

where
V_{rms} is the rms value of $v(t)$
I_{rms} is the rms value of $i(t)$

For the special case where both the current $i(t)$ and voltage $v(t)$ are sinusoidal, that is

$$v(t) = V_m \cos(wt + \theta_V) \tag{4.18}$$

and

$$i(t) = I_m \cos(wt + \theta_I) \tag{4.19}$$

the rms value of the voltage $v(t)$ is

$$V_{rms} = \frac{V_m}{\sqrt{2}} \tag{4.20}$$

and that of the current is

$$I_{rms} = \frac{I_m}{\sqrt{2}} \tag{4.21}$$

If a signal consists of both dc and ac components, as given by Equation 4.22

$$v_2(t) = V_{DC} + V_{m1} \cos(wt + \theta_V) \tag{4.22}$$

Then, the rms value of $v_2(t)$ is given as

$$V_{2rms} = \sqrt{V_{DC}^2 + \left(\frac{V_{m1}}{\sqrt{2}}\right)^2} \tag{4.23}$$

For triangular waveform centered about zero, the rms value is given as

$$V_{rms} = \frac{V_m}{\sqrt{3}} \tag{4.24}$$

In addition, for a square waveform centered about zero, the rms value is given as

$$V_{rms} = V_m \tag{4.25}$$

Example 4.2 shows how to use the Analog Discovery board to obtain the rms or effective value of periodic signals.

Example 4.2 Rms Values of Two Periodic Signals

Find the rms value of (1) a signal given as $v(t) = 1 + 2\cos(120\pi t)$ and (2) a triangular waveform with peak-to-peak value of 4V and zero voltage offset with frequency of 60Hz. (3) Compare your measured rms values with the calculated one.

Solution:

Build the Circuit:

Use a breadboard to build the circuit shown in Figure 4.7. Connect the AWG1 (yellow wire) to node 1 of the circuit. Also, connect scope channel 1 positive (orange wire) to node 1. In addition, connect the scope channel 1 negative (orange-white wire) to node 0. Moreover, connect the ground (black wire) to node 0 of the circuit. Furthermore, connect the AWG2 (yellow-white wire) to node 2 of the circuit, connect the scope channel 2 positive (blue wire) to node 2, and the scope channel 2 negative (blue-white wire) to node 0.

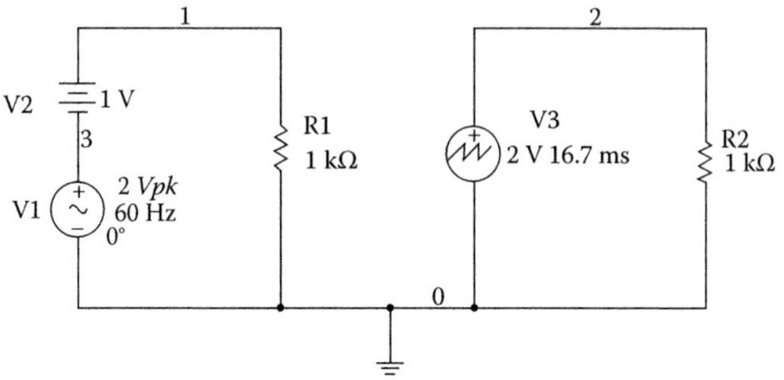

FIGURE 4.7
Circuit for performing rms measurements.

Generation of Sine and Square Waves:
Click on the WaveGen in the main WaveForms screen. Go to Basic → Sinusoid → Frequency (set to 60 Hz) → Amplitude (set to 2 V) → Offset (set to 1 V for dc voltage V2, as shown in Figure 4.7) → Symmetry (set to 50%) → Click Run AWG1.

Select Channel → Click on Channel 2 (AWG2) → Go to Basic → Triangular → Frequency (set to 60 Hz, which has a period of 16.7 ms) → Amplitude (set to 2 V) → Offset (set to 0 V) → Symmetry (set to 50%) → Click Run AWG2. Figure 4.8 shows the Analog Discovery WaveGen setup.

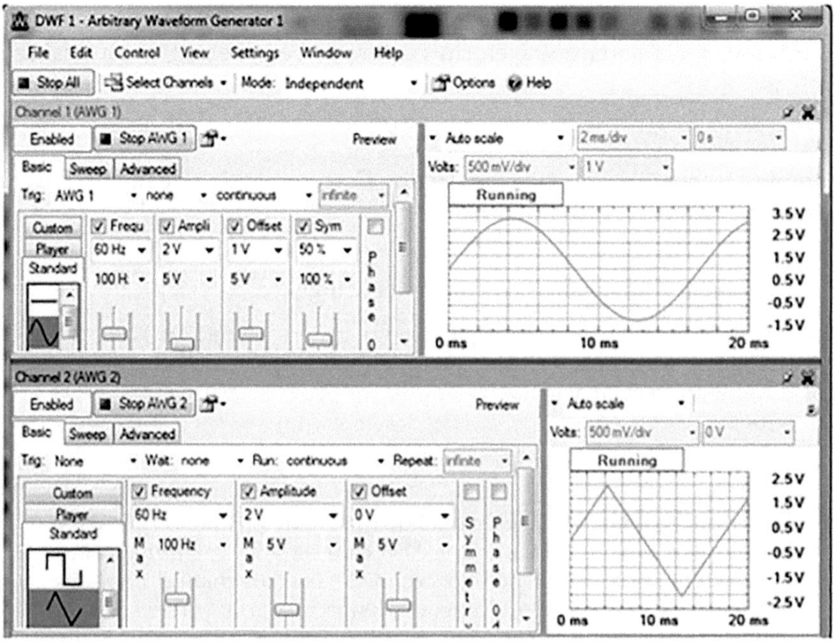

FIGURE 4.8
Analog Discovery WaveGen setup for sine and triangular waveforms.

Activation of Scope:
Click on the Scope in the main WaveForms screen. Go to Run → Autoset → Measure → Add → Channel 1 Vertical → dc rms → Add → Channel 1 Vertical → ac rms → Add Channel 1 Horizontal → Frequency → Add Selected Measurement → Close Add Instrument.

Measure → Add → Channel 2 Vertical → dc rms → Add → Channel 2 Vertical → ac rms → Add Selected Measurement → Close Add Instrument.

Figures 4.9 and 4.10 show the dc rms of (1) a sine wave with a dc component and (2) a triangular waveform, respectively. Table 4.1 captures

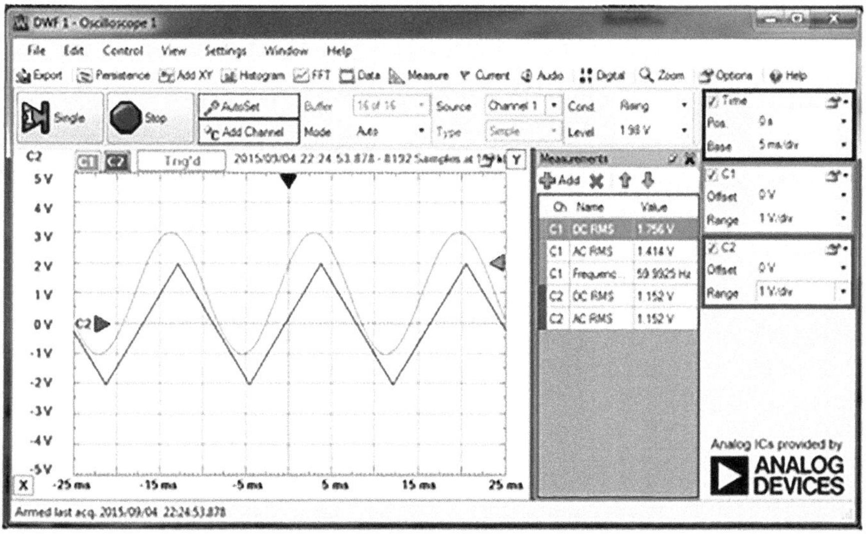

FIGURE 4.9
Display of (a) sine wave with a dc component (scope channel 1), and (b) triangular waveform (scope channel 2).

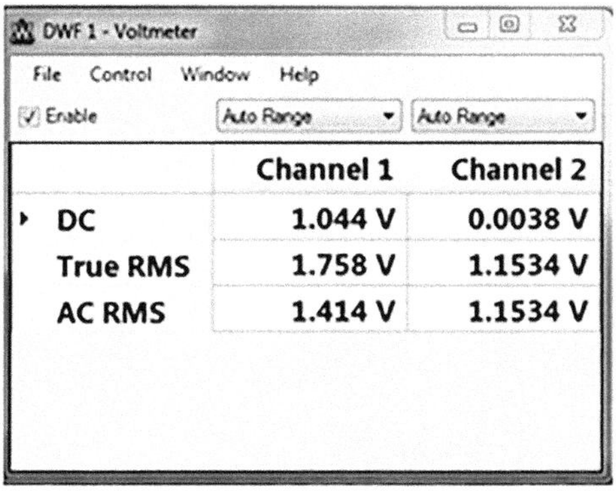

FIGURE 4.10
Voltmeter readings for (a) sine wave with dc voltage (Channel 1 of scope) and (b) triangular waveform (Channel 2 of scope).

TABLE 4.1
Measured rms Values of the Sinusoid with dc and the Triangular Wave

Signal	rms Values Calculated Using Equations 4.23 and 4.24	True rms Values Measured Using the Scope (Figure 4.9)	True rms Values Measured Using the Voltmeter (Figure 4.10)
Sine waveform with 1 V dc	1.732 V	1.758 V	1.758 V
Triangular waveform	1.1547 V	1.152 V	1.153 V

the information in Figures 4.9 and 4.10. It can be seen from the Table 4.1 that the measured values are very close to the theoretically calculated ones.

4.3 Complex Power

For the one-port network shown in Figure 4.1, the voltage across the network is given by $v(t)$ and current flowing into the network is $i(t)$. The instantaneous power $p(t)$ is

$$p(t) = v(t)i(t) \tag{4.26}$$

The average power dissipated by the one-port network is given as

$$P = \frac{1}{T}\int_0^T v(t)i(t)\,dt \tag{4.27}$$

The average power P is

$$P = V_{rms}I_{rms}\cos(\theta_V - \theta_I) \tag{4.28}$$

The power factor, pf, is given as

$$pf = \frac{P}{V_{rms}I_{rms}} \tag{4.29}$$

From Equations 4.28 and 4.29, we get

$$pf = \cos(\theta_V - \theta_I) \tag{4.30}$$

The reactive power Q is

$$Q = V_{rms} I_{rms} \sin(\theta_V - \theta_I) \tag{4.31}$$

and the complex power, S, is

$$S = P + jQ \tag{4.32}$$

$$S = V_{rms} I_{rms} \left[\cos(\theta_V - \theta_I) + j \sin(\theta_V - \theta_I) \right] \tag{4.33}$$

The magnitudes V_m and I_m can be obtained from the scope, and the phase difference $\theta_Z = (\theta_V - \theta_I)$ can also be determined by measuring the phase difference between the voltage and current waveforms.

If t_d is the phase difference between the voltage and current waveforms, and T_P is the period of the voltage or current waveform, then the phase difference is given by Equation 4.9. The steps that can used to obtain the phase difference between the voltage and current waveforms have been described in Section 4.1.

To measure the current, a known value resistor will be placed in series with the circuit under test. Channel 1 of the scope will be connected across the known resistor, and channel 2 will be connected across the circuit under test. The Mathematic Channel can be used to monitor the current by dividing the voltage across the known valued resistor by the resistance of the resistor. Example 4.3 describes how to use the Analog Discovery board to obtain the complex power of a circuit.

Example 4.3 Determining the Complex Power of a Circuit

For the circuit shown in Figure 4.11, (1) find the complex power supplied by the source. (2) If 1.0 μF capacitor is connected across the combined 200 Ω resistor and the 40 mH inductor, what is the complex power supplied by the source? (3) Compare the power factor of part (1) to that of part (2).

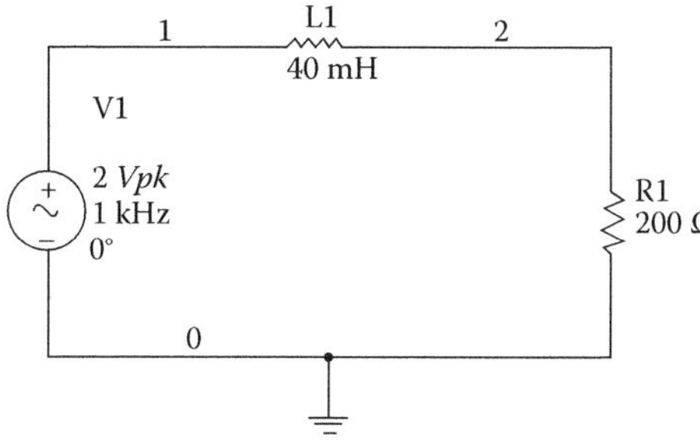

FIGURE 4.11
Complex power for RL circuit.

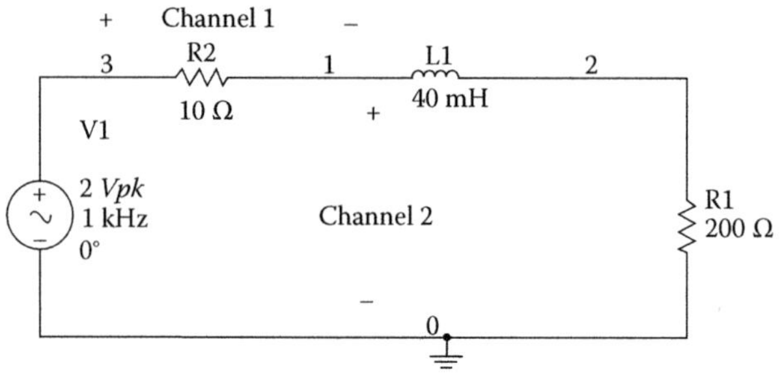

FIGURE 4.12
Augmented version of Figure 4.11 showing 10 Ω resistor to monitor current.

Solution:
A 10 Ω resistor has been inserted in series with the circuit to measure the current flowing into the RL circuit. The augmented circuit is shown in Figure 4.12.

Build the Circuit:
Use a breadboard to build the circuit shown in Figure 4.12. Connect the AWG1 (yellow wire) to node 3 of the circuit. Also, connect the scope channel 1 positive (orange wire) to node 3. In addition, connect the scope channel 1 negative (orange-white wire) to node 1. Moreover, connect the ground (black wire) to node 0 of the circuit. Furthermore, connect the scope channel 2 positive (blue wire) to node 1, and the scope channel 2 negative (blue-white wire) to node 0.

Channel 1 measures the voltage across the 10 Ω resistor, and channel 2 measures the voltage at the input of the series RL circuit.

Generation of Sine Wave:
Click on the WaveGen in the main WaveForms screen. Go to Basic → Sinusoid → Frequency (set to 1 kHz) → Amplitude (set to 4 V) → Offset (set to 0 V) → Symmetry (set to 50%) → Click Run AWG1. Figure 4.13 shows the Analog Discovery WaveGen setup.

Activation of Scope:
Click on the Scope in the main WaveForms screen. Go to Run → Autoset → Measure → Add → Channel 1 Horizontal → Period → Add → Channel 1 Vertical → rms voltage → Add → Channel 2 Vertical → rms voltage → Control → Add Mathematic Channel → Custom → C1/10 → Add → Channel M1 Vertical → rms voltage → Change Math Mode Units to A → Add Selected Measurement → Close Add Instrument.

From Figure 4.14, we get the following measurements:

Rms voltage at the input of the RL circuit at nodes 1 and 0 of Figure 4.12 is 1.386 V.

Rms value of the current flowing through the 10 Ω resistor of Figure 4.12 is 3.639 mA.

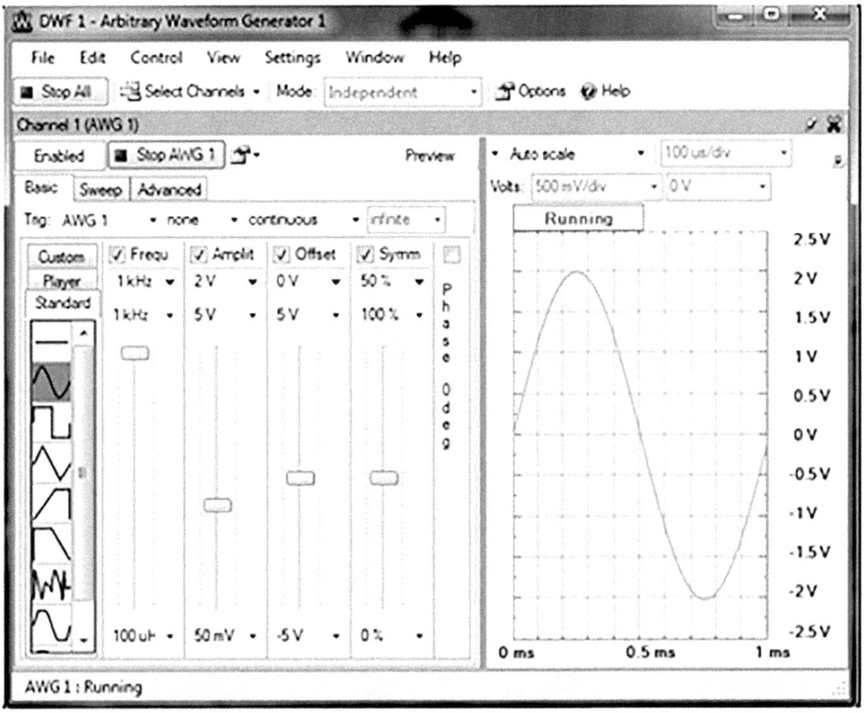

FIGURE 4.13
Analog Discovery WaveGen setup for complex power measurement.

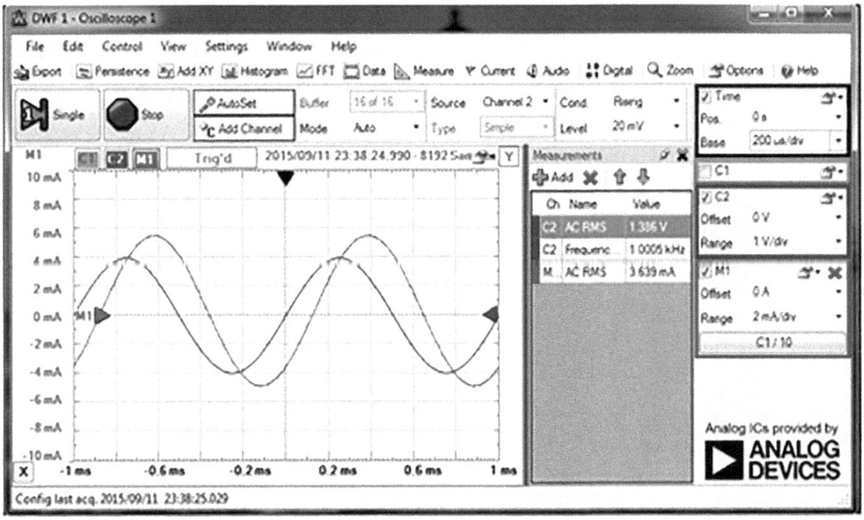

FIGURE 4.14
Scope display showing the voltage (scope channel 2) and current (mathematic channel 1) waveforms.

Measurements made on the scope display show that the time difference between the voltage and current is **124 μs.** Using Equation 4.10, we have

$$|\theta_V - \theta_I| = \frac{t_d}{T_P}(360°) = \frac{124\ \mu s}{1\ ms} \times 360° = 44.64° \tag{4.34}$$

From Figure 4.14, the voltage waveform (C2) attains its maximum value earlier than that of the current waveform (M1). Therefore, the voltage leads the current. Thus

$$\theta_Z = (\theta_V - \theta_I) = 44.64° \tag{4.35}$$

Using Equation 4.33, the complex power is given as

$$S = (1.386)(0.003639)\left[\cos(44.64°) + j\sin(44.64°)\right]\ \text{VA}$$

$$S = 3.5898 + j3.544\ \text{mVA} \tag{4.36}$$

When the capacitor is connected, we have the circuit shown in Figure 4.15.

The display at channel 1 and channel 2 of the scope is shown in Figure 4.16. From Figure 4.16, we get the following measurements:

Rms voltage at the input of RL circuit at nodes 1 and 0 of Figure 4.16 is 1.3796 V.

Rms value of the current flowing through the 10 Ω resistor of Figure 4.16 is 5.85 mA.

Measurements made on the scope display show that the time difference between the voltage and current is **164 μs.** Using Equation 4.10, we have

$$|\theta_V - \theta_I| = \frac{t_d}{T_P}(360°) = \frac{164\ \mu s}{1\ ms} \times 360° = 59.04° \tag{4.37}$$

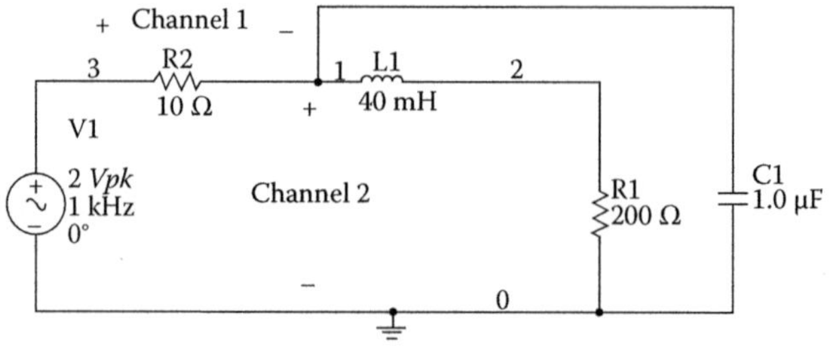

FIGURE 4.15
RL circuit with capacitor 1.0 μF connected across both RL elements.

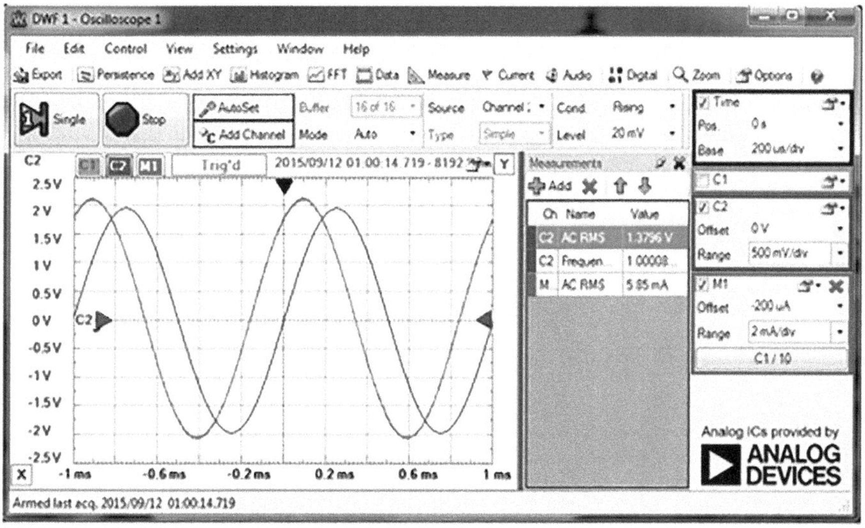

FIGURE 4.16
Display of current (mathematic channel 1) and voltage (scope channel 2) waveforms with capacitor connected across RL elements.

From Figure 4.16, the current waveform (M1) attains its maximum value earlier than that of the voltage waveform (C2). Therefore, the voltage lags the current. Thus

$$\theta_Z = (\theta_V - \theta_I) = -59.04^\circ \tag{4.38}$$

Using Equation 4.33, the complex power is given as

$$S = (1.3796)(0.00585)\left[\cos(-59.04^\circ) + j\sin(-59.04^\circ)\right] \text{ VA}$$

$$S = 4.1498 - j6.9173 \text{ mVA} \tag{4.39}$$

From Equation 4.30, the power factor before the capacitor was connected is given as

$$pf = \cos(\theta_V - \theta_I) = \cos(44.64^\circ) = 0.7115 \text{ lagging} \tag{4.40}$$

Using Equation 4.30 again, the power factor after the capacitor was connected across both the resistor and inductor is given as

$$pf = \cos(\theta_V - \theta_I) = \cos(-59.04^\circ) = 0.5144 \text{ leading} \tag{4.41}$$

4.4 Frequency Response

Figure 4.17 shows a linear network with input $x(t)$ and output $y(t)$. Its complex frequency representation is also shown. In general, the input $x(t)$ and output $y(t)$ are related by the differential equation

$$a_n \frac{d^n y(t)}{dt^n} + a_{n-1}\frac{d^{n-1}y(t)}{dt^{n-1}} + \cdots + a_1 \frac{dy(t)}{dt} + a_0 y(t)$$
$$= b_m \frac{d^m x(t)}{dt^m} + b_{m-1}\frac{d^{m-1}x(t)}{dt^{m-1}} + \cdots + b_1 \frac{dx(t)}{dt} + b_0 x(t) \tag{4.42}$$

where $a_n, a_{n-1}, \ldots, a_0, b_m, b_{m-1}, \ldots, b_0$ are real constants.

If $x(t) = X(s)e^{st}$, then the output must have the form $y(t) = Y(s)e^{st}$, where $X(s)$ and $Y(s)$ are phasor representations of $x(t)$ and $y(t)$. From Equation 4.42, we have

$$(a_n s^n + a_{n-1}s^{n-1} + \cdots + a_1 s + a_0)Y(s)e^{st}$$
$$= (b_m s^m + b_{m-1}s^{m-1} + \cdots + b_1 s + b_0)X(s)e^{st} \tag{4.43}$$

and the network function

$$H(s) = \frac{Y(s)}{X(s)} = \frac{b_m s^m + b_{m-1}s^{m-1} + \cdots b_1 s + b_0}{a_n s^n + a_{n-1}s^{n-1} + \cdots a_1 s + a_0} \tag{4.44}$$

More specifically, for a second-order analog filter, the following transfer functions can be obtained:

1. Lowpass

$$H_{LP}(s) = \frac{k_1}{s^2 + Bs + w_0^2} \tag{4.45}$$

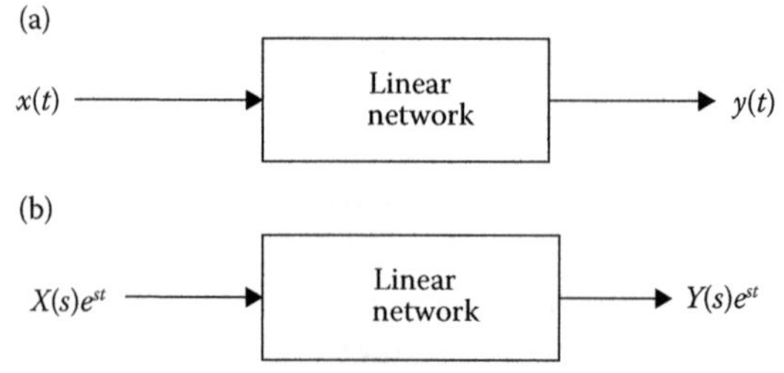

FIGURE 4.17
Linear network representation of (a) time domain and (b) s-domain.

2. Highpass

$$H_{HP}(s) = \frac{k_2 s^2}{s^2 + Bs + w_0^2} \tag{4.46}$$

3. Bandpass

$$H_{BP}(s) = \frac{k_3 s}{s^2 + Bs + w_0^2} \tag{4.47}$$

4. Band reject

$$H_{BR}(s) = \frac{k_4 s^2 + k_5}{s^2 + Bs + w_0^2} \tag{4.48}$$

where k_1, k_2, k_3, k_4, B, and w_0 are constants.

Frequency response is the response of a network to sinusoidal input signal. If we substitute $s = jw$ in the general network function, $H(s)$, we get

$$H(s)\Big|_{s=jw} = M(w)\angle\theta(w) \tag{4.49}$$

where

$$M(w) = |H(jw)| \tag{4.50}$$

and

$$\theta(w) = \angle H(jw) \tag{4.51}$$

The plot of $M(\omega)$ versus ω is the magnitude characteristics or response. Also, the plot of $\theta(w)$ versus ω is the phase response. The magnitude and phase characteristics can be obtained by using the Analog Discovery board. The Network Analyzer of the Analog Discovery board can be used to plot the magnitude and phase characteristics of a circuit (see Section 1.8). Example 4.4 uses the Network Analyzer of the Analog Discovery board to obtain the frequency response of a RLC resonant circuit.

Example 4.4 Frequency Response of a RLC Resonant Circuit

For the circuit shown in Figure 4.18, (1) use the Network Analyzer to plot the magnitude and phase characteristics. (2) Determine the resonant frequency and bandwidth of the circuit. (3) Calculate the theoretical values of the resonant frequency and the bandwidth of the series RLC circuit. (4) Compare the results obtained in parts (2) and (3).

Solution:

Build the Circuit:

Use a breadboard to build the circuit shown in Figure 4.18. Connect the AWG1 (yellow wire) to node 1 of the circuit. Also, connect the scope

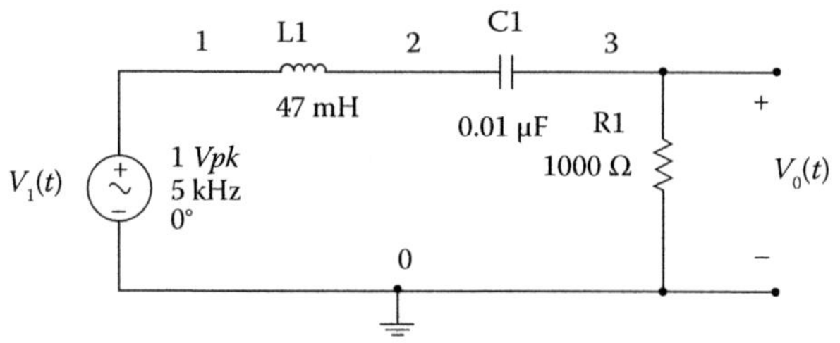

FIGURE 4.18
RLC resonant circuit.

channel 1 positive (orange wire) to node 1. In addition, connect the scope channel 1 negative (orange-white wire) to node 0. Moreover, connect the ground (black wire) to node 0 of the circuit. Furthermore, connect the scope channel 2 positive (blue wire) to node 3, and the scope channel 2 negative (blue-white wire) to node 0.

Generation of Sine Wave:
Click on the WaveGen in the main WaveForms screen. Go to Basic → Sinusoid → Frequency (set to 5 kHz) → Amplitude (set to 1 V) → Offset (set to 0 V) → Symmetry (set to 50%) → Click Run AWG1. Figure 4.19 shows the Analog Discovery WaveGen setup.

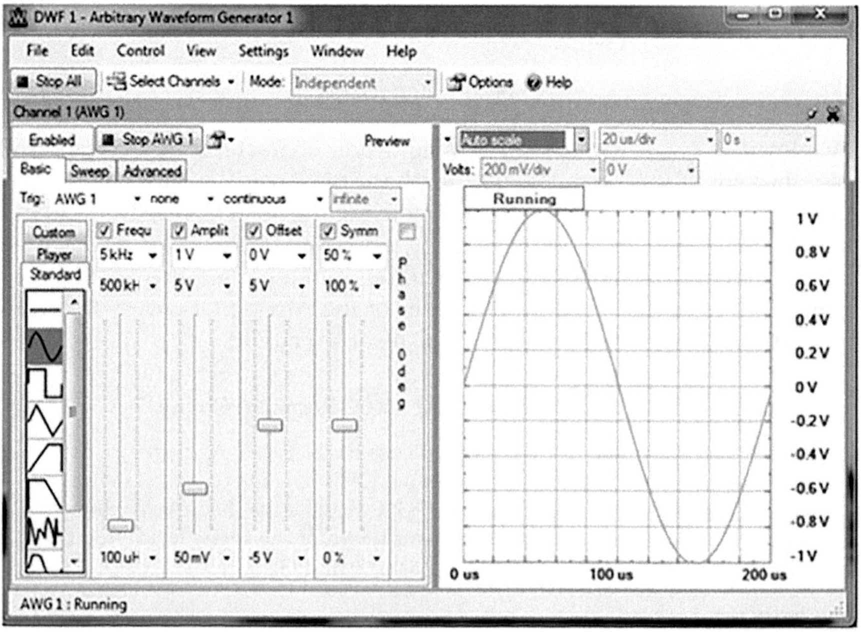

FIGURE 4.19
Analog Discovery WaveGen setup for obtaining frequency response.

Activation of Network Analyzer:
The Network Analyzer uses both oscilloscope channels as the input and output channels. Channel 1 is for measuring the input signal, and channel 2 is for measuring the output signal. Open up the Network Analyzer that can be found under the More Instruments tab. Use the following setup values:

Start Frequency: 1 kHz
Stop Frequency: 50 kHz
Offset: 0 V
Input Signal Amplitude: 1 V
Max Filter Gain: 1X
Bode Scale: Magnitude—Top 5 dB, Range of 40 dB
Phase—Top 90°, Range of 180°
Scope Channel Gain: Channel 1: 1X; Channel 2: 1X
Obtain a single sweep in frequency by clicking Single. This provides a Bode plot representation of the frequency response of the circuit. The Bode plot obtained is shown in Figure 4.20.

From Figure 4.20, the following measurements were obtained:

Resonant frequency is 8.76 kHz.

Low cut-off frequency is 6.9 kHz.

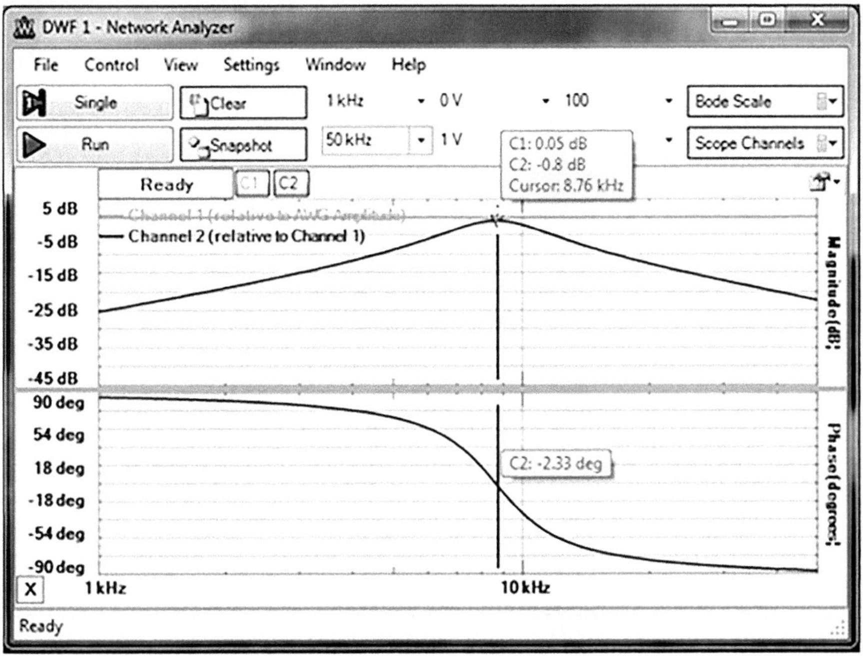

FIGURE 4.20
Magnitude and phase response of the RLC circuit of Figure 4.18.

TABLE 4.2

Resonant Frequency, Quality Factor, and Bandwidth of RLC Circuit

Values	Resonant Frequency (kHz)	Quality Factor	Bandwidth (kHz)
Measured value	8.76	2.16	4.06
Calculated value	7.34	2.17	3.39

High cut-off frequency is 10.96 kHz.

The bandwidth = 8.76 − 6.9 kHz = 4.06 kHz.

The quality factor is $\frac{8.76}{4.06} = 2.16$.

For the series RLC circuit, the resonant frequency ω_0 is given as

$$\omega_0 = \frac{1}{\sqrt{LC}} \tag{4.52}$$

The quality factor Q is

$$Q = \frac{\omega_0 L}{R} \tag{4.53}$$

Bandwidth BW is given as

$$BW = \frac{\omega_0}{Q} \tag{4.54}$$

Using Equations 4.52, 4.53, and 4.54, respectively, one can calculate the resonant frequency, quality factor, and bandwidth. Table 4.2 shows the comparison between the measured and calculated values:

PROBLEMS

Problem 4.1 Use the Analog Discovery board to determine the input impedance of the RL circuit shown in Figure P4.1 between the nodes 1 and 0. Compare your measured impedance with the calculated value of the input impedance.

Problem 4.2 Use the Analog Discovery board to find the input impedance of the RC circuit shown in Figure P4.2 between the nodes 1 and 0. Compare your measured impedance with the calculated value of the input impedance.

Problem 4.3 Use the Analog Discovery board to determine the input impedance of the RC circuit shown in Figure P4.3 between the nodes 1 and 0. Compare your measured impedance with the calculated value of the input impedance.

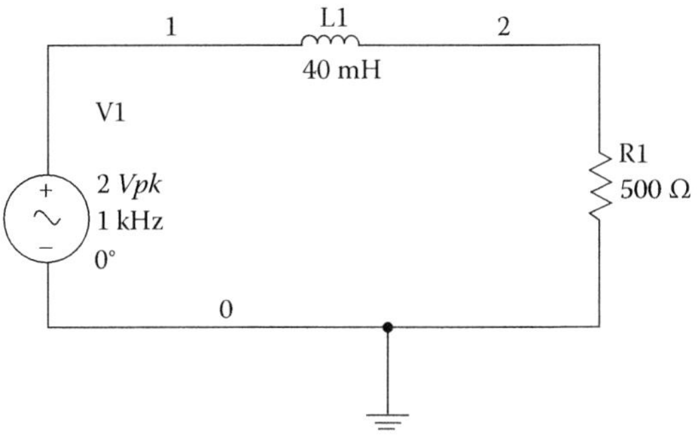

FIGURE P4.1
RL circuit.

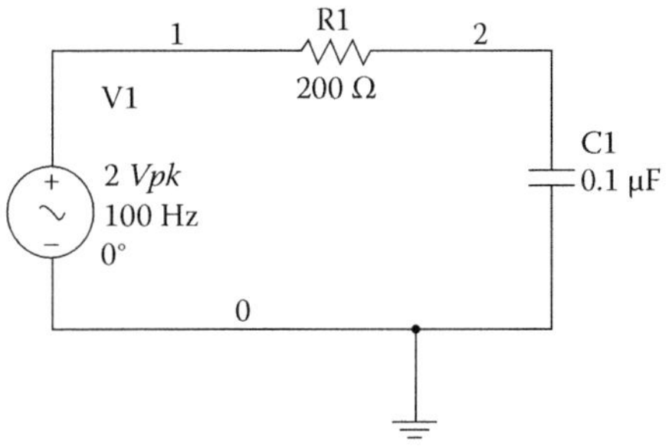

FIGURE P4.2
RC circuit.

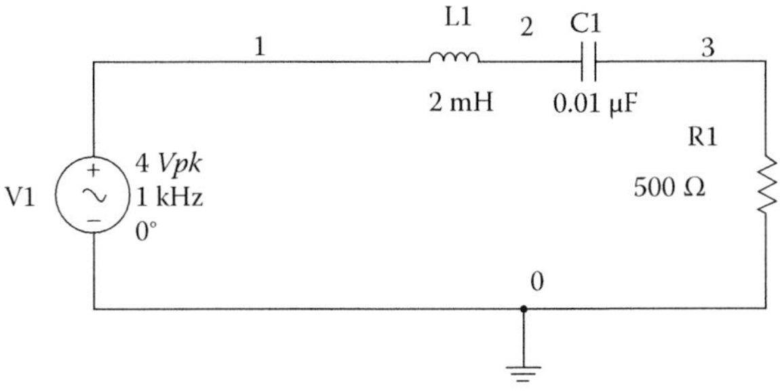

FIGURE P4.3
RLC circuit.

Problem 4.4 Use the Analog Discovery board to find the rms value of the signal given as

$$V_{p4}(t) = 2 + 3\cos(377t) \text{ V}$$

Compare the experimentally measured value of the rms with the calculated value.

Problem 4.5 Use the Analog Discovery board to find the rms value of a square wave with peak-to-peak value of 6 V and dc value of 2 V. Assume that the frequency of the square wave is 2 kHz. The waveform is shown in Figure P4.5.

Problem 4.6 (1) Use the Analog Discovery board to find the rms value of the triangular waveform of frequency 2 kHz, shown in Figure P4.6. (2) Compare

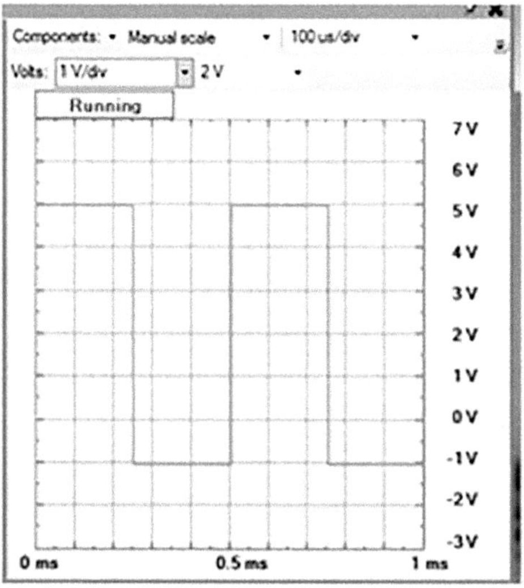

FIGURE P4.5
Square waveform with dc offset value.

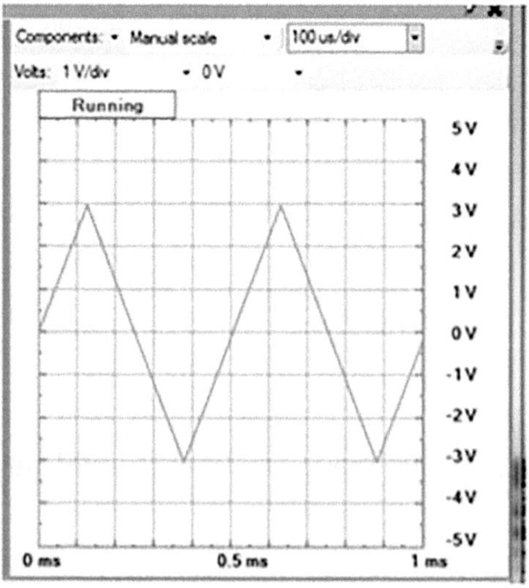

FIGURE P4.6
Triangular waveform.

the rms value obtained experimentally with that obtained theoretically. (3) If the waveform is assumed to be a voltage across 10 Ω resistor, what will be the average power dissipated in the resistor?

Problem 4.7 (1) For circuit shown in Figure P4.7, find the complex power supplied by the source with the use of the Analog Discovery board. (2) If a 1.0 μF capacitor is disconnected across the combined 100 Ω resistor and the 2 mH inductor, what is the complex power supplied by the source? (3) Compare the power factor of part (1) to that of part (2).

Problem 4.8 (1) For the circuit shown in Figure P4.8, find the complex power supplied by the source. (2) If a 1.0 μF capacitor is disconnected across the 100 Ω resistor, what is the complex power supplied by the source? (3) Compare the power factor of part (1) to that of part (2).

Problem 4.9 For the circuit shown in Figure P4.9, (1) use the Network Analyzer to plot the magnitude and phase characteristics. (2) Determine the resonant frequency and bandwidth of the circuit. (3) Compare the measured values with the theoretically calculated values for the resonant frequency and the bandwidth of the circuit.

Problem 4.10 For the circuit shown in Figure P4.10, (1) use the Network Analyzer to plot the magnitude and phase characteristics. (2) Determine the resonant frequency and bandwidth of the circuit.

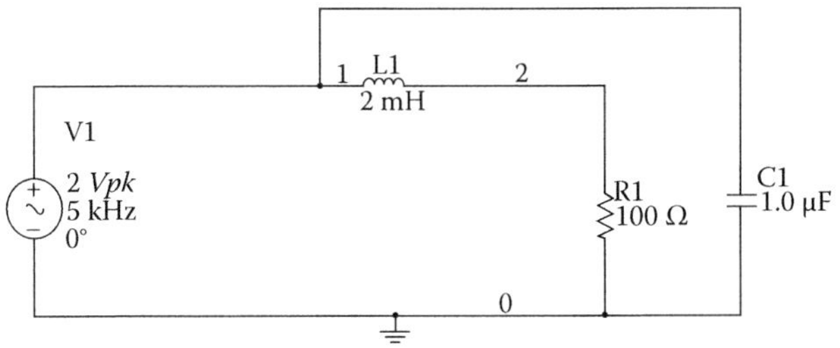

FIGURE P4.7
RLC circuit for complex power calculations.

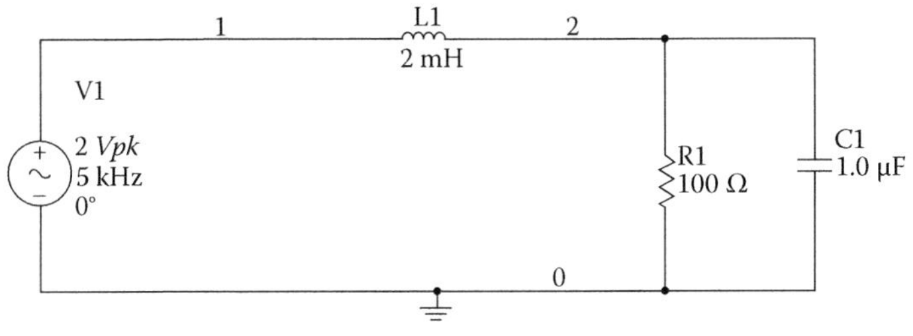

FIGURE P4.8
RLC circuit for power calculations.

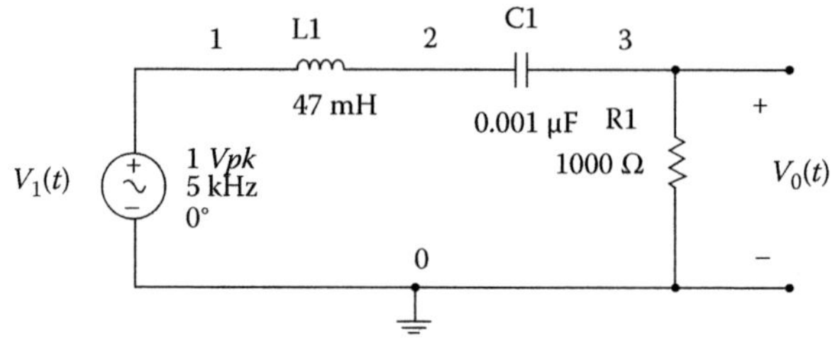

FIGURE P4.9
Resonant circuit.

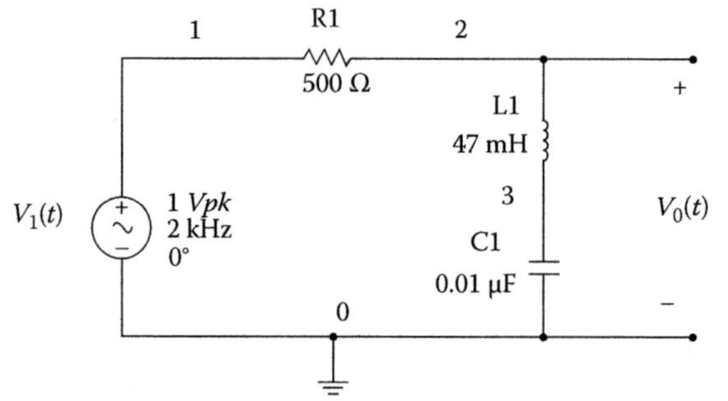

FIGURE P4.10
Notch filter circuit.

5

Operational Amplifiers

INTRODUCTION The operational amplifier (op amp) is one of the most versatile electronic circuits. It can be used to perform the basic mathematical operations: addition, subtraction, multiplication, and division. It can also be used to do integration and differentiation. Several electronic circuits use an op amp as an integral element; some of these circuits are amplifiers, filters, and oscillators. In this chapter, the voltage gains of inverting amplifiers, non-inverting amplifiers and weighted summers will be measured with the Analog Discovery board. In addition, integrators, differentiators, and active filters will be discussed with worked examples.

5.1 Properties of the Op Amp

The op amp, from a signal point of view, is a three-terminal device: two inputs and one output. Its symbol is shown in Figure 5.1. The inverting input is designated by the "–" sign and non-inverting input by the "+" sign.

An ideal op amp has the following properties:

- infinite input resistance
- zero output resistance
- zero offset voltage
- infinite frequency response
- infinite common-mode rejection ratio
- infinite open-loop gain, A.

A practical op amp will have a large but finite open-loop gain in the range from 10^5 to 10^9. It also has a very large input resistance, 10^6 to 10^{10} Ω. The output resistance might be in the range of 50 to 125 Ω. The offset voltage is small but finite, and the frequency response will deviate considerably from the infinite frequency response. The common-mode rejection ratio is not infinite but finite. Table 5.1 shows the properties of two general purpose operational amplifiers.

The pin diagrams for LM741 (from www.ti.com) and OP484 (from www.analog.com) operational amplifiers are shown in Figures 5.2 and 5.3, respectively.

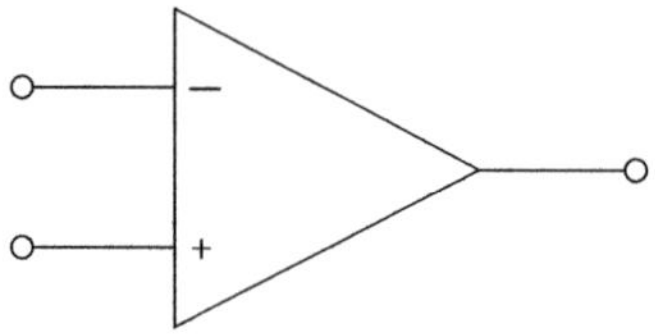

FIGURE 5.1
Op amp circuit symbol.

TABLE 5.1

Properties of LM741 and OP484 Op Amps

Property	Typical Value for LM741 Op Amp	Typical Value for OP484 Op Amp
Open-loop gain	2.0×10^5	1.5×10^5
Offset voltage	1 mV	60 μV
Input bias current	30 nA	80 nA
Unity-gain bandwidth	1 MHz	4.25 MHz
Common-mode rejection ratio	95 dB	90 dB
Slew rate	0.7 V/μV	4.0 V/μV

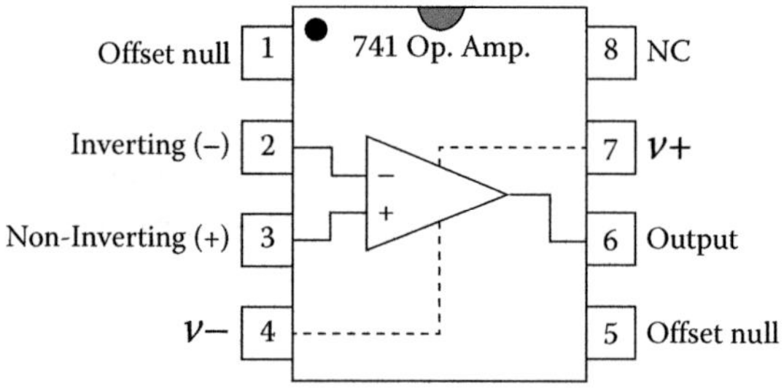

FIGURE 5.2
Pin diagram of operational amplifier 741.

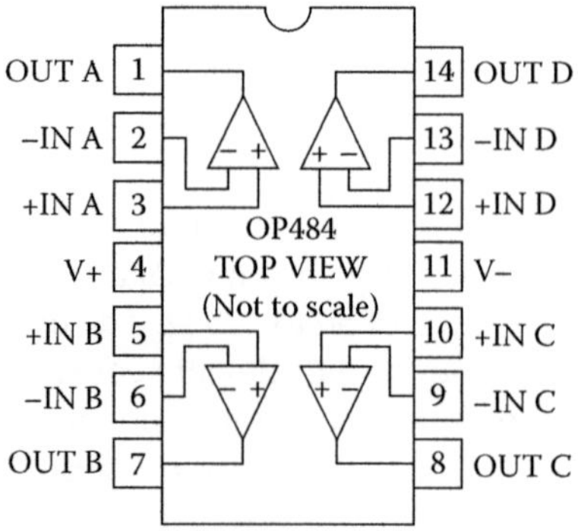

FIGURE 5.3
Pin diagram of operational amplifier OP484.

5.2 Inverting Amplifier

An op amp circuit connected in an inverted closed-loop configuration is shown in Figure 5.4.

It can be shown that the closed-loop gain of the amplifier is

$$\frac{V_O}{V_{IN}} = -\frac{R_2}{R_1} \tag{5.1}$$

and the input resistance is R_1. Normally, $R_2 > R_1$ such that $|V_0| > |V_{in}|$. Example 5.1 shows how to use the Analog Discovery board to obtain the voltage gain of a closed-loop inverting amplifier.

Example 5.1 Inverting Amplifier

For the op amp circuit shown in Figure 5.5, (1) use the Analog Discovery board to obtain the gain of the amplifier. (2) Compare the measured gain with the theoretically calculated gain.

Solution:

Build the Circuit:

Use a breadboard to build the circuit shown in Figure 5.5. Connect the AWG1 (yellow wire) to node 1 of the circuit. Also, connect the scope

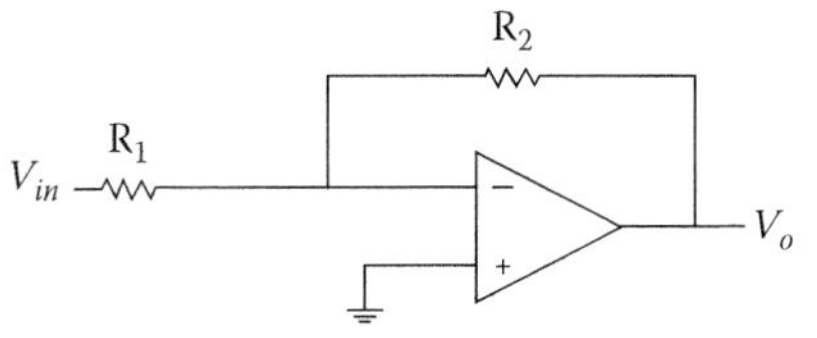

FIGURE 5.4
Inverting amplifier.

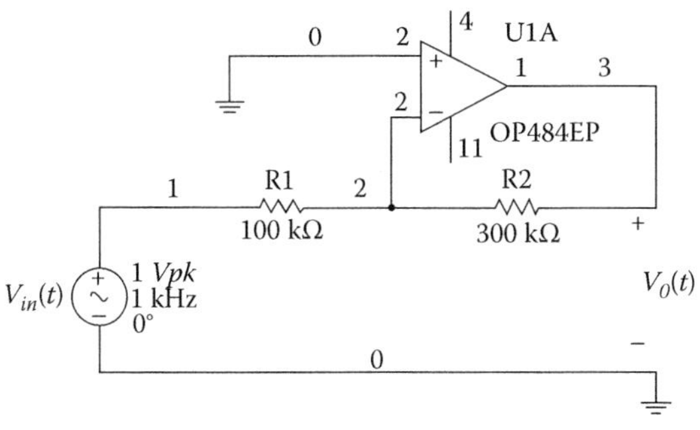

FIGURE 5.5
Inverting amplifier built using OP484EP Op amp.

channel 1 positive (orange wire) to node 1. In addition, connect the scope channel 1 negative (orange-white wire) to node 0. Moreover, connect the ground (black wire) to node 0 of the circuit. Furthermore, connect the scope channel 2 positive (blue wire) to node 3, and the scope channel 2 negative (blue-white wire) to node 0.

For the op amp OP484, connect the inverting input (pin 2) to node 2 of the circuit. In addition, connect the non-inverting input (pin #3) to node 0. Moreover, connect the op amp output (pin #1) to node 3. Furthermore, connect op amp pin #4 to the 5 V voltage source (red wire of the Analog Discovery). Furthermore, connect the op amp pin #11 to the –5 V voltage source (white wire of the Analog Discovery).

Generation of Sine Wave:
Click on the WaveGen in the main WaveForms screen. Go to Basic → Sine → Frequency (set to 1000 Hz) → Amplitude (set to 1 V) → Offset (set to 0 V) → Symmetry (set to 50%) → Click Run AWG1. Figure 5.6 shows the Analog Discovery WaveGen setup.

Activation of the Power Supply:
Click Voltage in the main WaveForms screen → Click Power to turn on the voltage sources → Click V+ to turn on the 5 V source → Click V– to turn on the –5 V voltage source → Upon completion of the measurements, click Power again to turn it off. Figure 5.7 shows the +5 V and –5 V voltage sources used for powering the op amp.

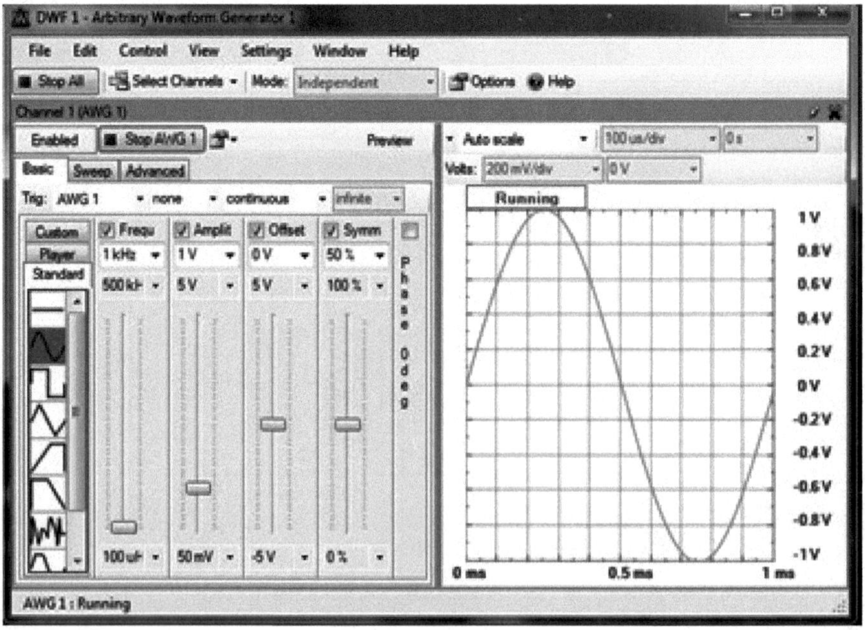

FIGURE 5.6
Sinusoidal waveform produced by WaveGen.

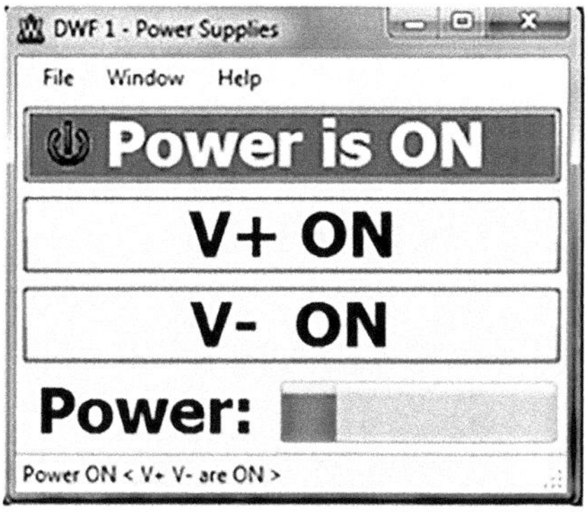

FIGURE 5.7
5 V and −5 V analog discovery power supplied to the inverting amplifier.

> *Activation of Scope:*
> Click on the Scope in the main WaveForms screen. Go to Run → Autoset → Measure → Add → Channel 1 → Horizontal → Frequency → Add Selected Measurement → Add → Channel 1 → Vertical → Peak to Peak → Add Selected Instrument → Add → Channel 2 → Vertical → Peak to Peak → Add Selected Measurement → Close Add Instrument.

Figure 5.8 shows the input and output waveforms displayed on the scope.

From Figure 5.8, it can be seen that the input and output waveforms are out of phase by 180°. In addition, from Figure 5.8, one can obtain the following voltages:

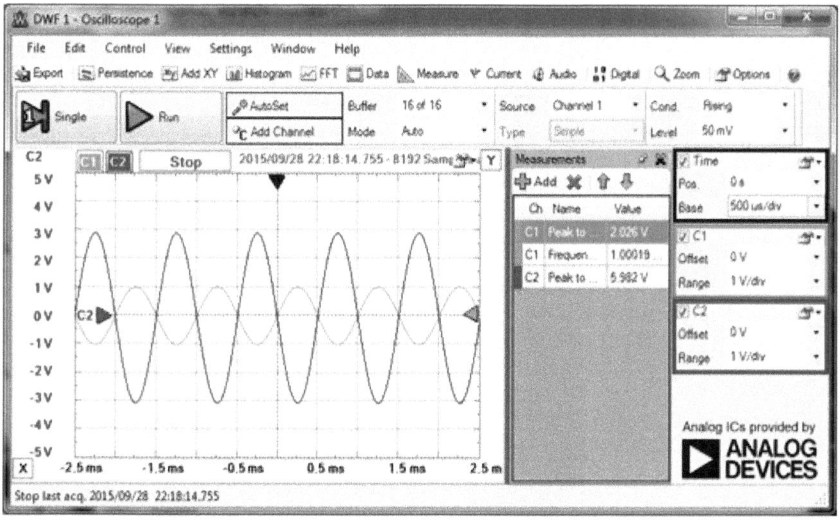

FIGURE 5.8
Display of input (scope channel 1) and output (scope channel 2) sinusoidal waveforms of an inverting amplifier.

Peak-to-peak voltage of the input signal=2.026 V.

Peak-to-peak voltage of the output signal=5.982 V.

The voltage gain obtained from scope measurements is given as

$$\frac{V_O}{V_{in}} = -\frac{5.982}{2.026} = -2.95 \tag{5.2}$$

From Equation 5.1, the theoretical gain of the inverting amplifier is given as

$$\frac{V_O}{V_{in}} = -\frac{R_2}{R_1} = -\frac{300\text{ K}}{100\text{ K}} = -3 \tag{5.3}$$

From Equations 5.2 and 5.3, it can be seen that the measured gain is very close to the theoretically calculated gain.

5.3 Non-inverting Amplifier

An op amp circuit connected in a non-inverting closed-loop configuration is shown in Figure 5.9. It can be shown that the voltage gain is

$$\frac{V_O}{V_{in}} = \left(1 + \frac{R_2}{R_1}\right) \tag{5.4}$$

The gain of the non-inverting amplifier is positive, meaning that the waveforms at the input and output of the circuit are in phase. The input impedance of the amplifier approaches infinity, since the current that flows into the positive input of the op amp is almost zero. Example 5.2 shows how to use the Analog Discovery board to obtain the voltage gain of a non-inverting amplifier.

Example 5.2 Non-inverting Amplifier

For the op amp circuit shown in Figure 5.10, (1) use the Analog Discovery board to obtain the gain of the amplifier. (2) Compare the measured voltage gain with the theoretically calculated gain.

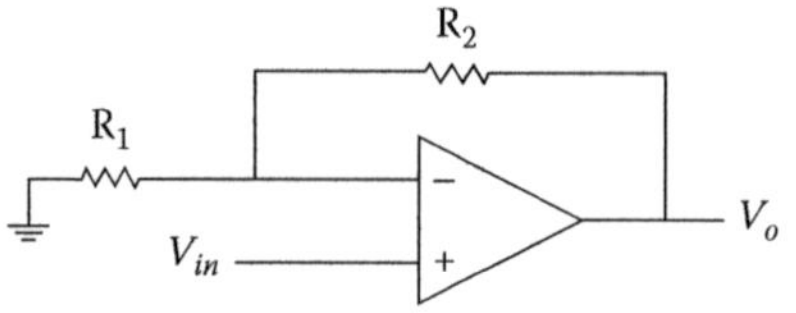

FIGURE 5.9
Non-inverting amplifier.

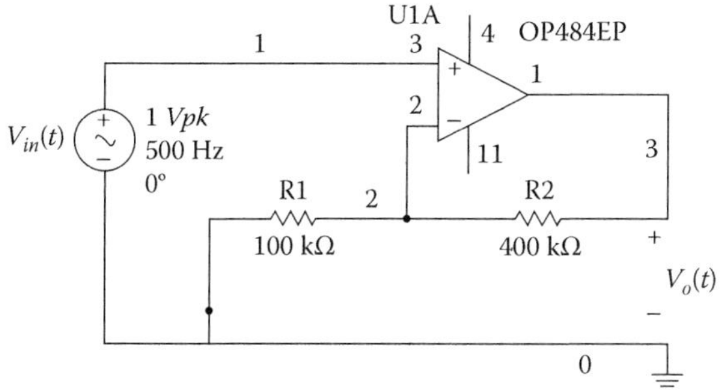

FIGURE 5.10
Non-inverting amplifier built using OP484EP op amp.

Solution:
Build the Circuit:
Use a breadboard to build the circuit shown in Figure 5.10. Connect the AWG1 (yellow wire) to node 1 of the circuit. Also, connect the scope channel 1 positive (orange wire) to node 1. In addition, connect the scope channel 1 negative (orange-white wire) to node 0. Furthermore, connect the ground (black wire) to node 0 of the circuit. Moreover, connect the scope channel 2 positive (blue wire) to node 3, and the scope channel 2 negative (blue-white wire) to node 0.

For the op amp, connect the inverting input (pin #2) to node 2 of the circuit. In addition, connect the non-inverting input (pin #3) to node 1. Moreover, connect the op amp output (pin #1) to node 3. Furthermore, connect the op amp pin #4 to the 5 V voltage source (red wire of the Analog Discovery). Furthermore, connect the op amp pin #11 to the –5 V voltage source (white wire of the Analog Discovery).

Generation of Sine Wave:
Click on the WaveGen in the main WaveForms screen. Go to Basic → Sine → Frequency (set to 1000 Hz) → Amplitude (set to 1 V) → Offset (set to 0 V) → Symmetry (set to 50%) → Click Run AWG1. Figure 5.11 shows the Analog Discovery WaveGen setup for generating a sinusoidal waveform.

Activation of the Power Supply:
Click Voltage in the main WaveForms screen → Click Power to turn on the voltage sources → Click V+ to turn on the 5 V source → Click V– to turn on the –5 V voltage source. Upon completion of the measurements, click Power again to turn it off. Figure 5.12 shows the +V and –V voltage sources for supplying power to the op amp.

Activation of Scope:
Click on the Scope in the main WaveForms screen. Go to Run → Autoset → Measure → Add → Channel 1 → Horizontal → Frequency → Add Selected Measurement → Add → Channel → Vertical → Peak to Peak → Add Selected Instrument → Add → Channel 2 → Vertical → Peak to Peak → Add Selected Measurement → Close Add Instrument. Figure 5.13 shows the input and output waveforms displayed on the scope.

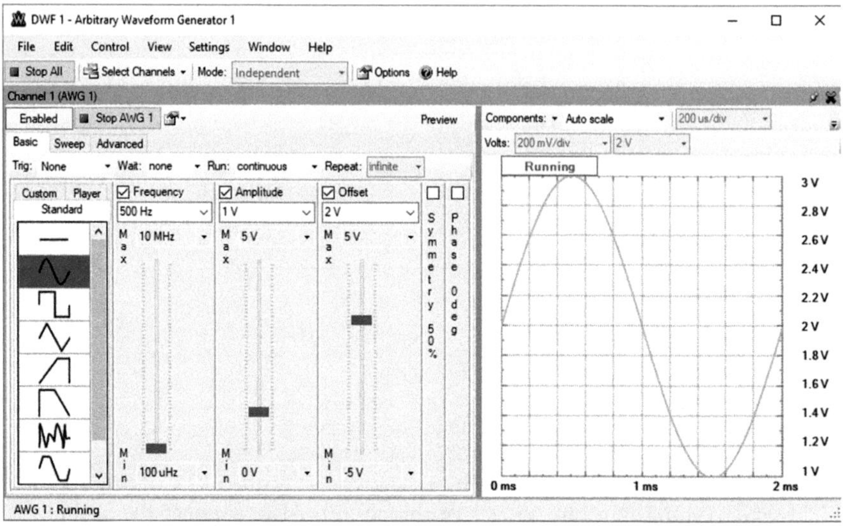

FIGURE 5.11
Sinusoidal waveform produced by WaveGen AWG1.

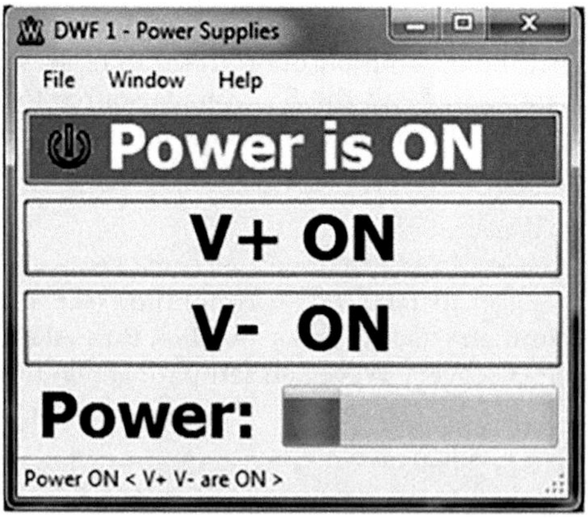

FIGURE 5.12
5 V and –5 V voltage sources supplied to the non-inverting amplifier.

From Figure 5.13, it can be seen that the input and output waveforms are in phase. In addition, from Figure 5.13, we obtain the following voltages:

Peak-to-peak voltage of the input signal is 2.02 V.

Peak-to-peak voltage of the output signal is 9.954 V.

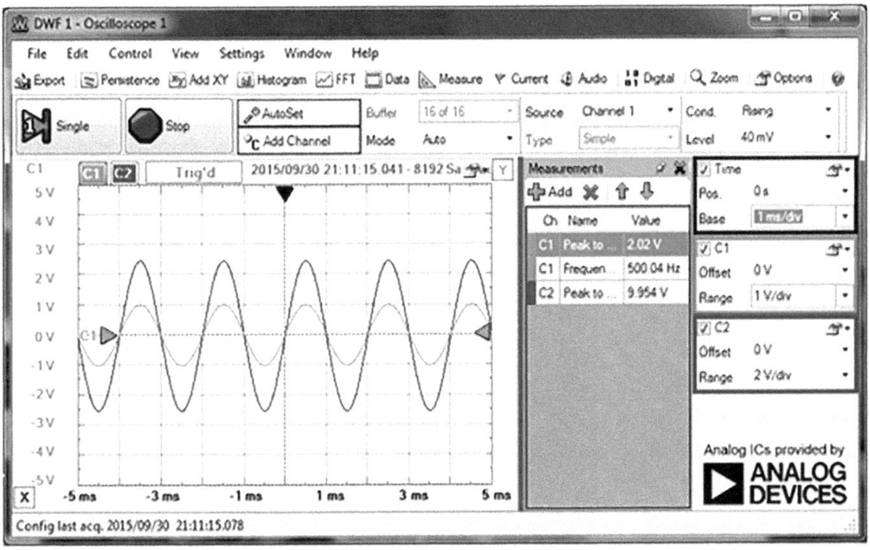

FIGURE 5.13
Display of input (scope channel 1) and output (scope channel 2) sinusoidal waveforms of the non-inverting amplifier.

The voltage gain obtained from the scope measurements is given as

$$\frac{V_O}{V_{IN}} = \frac{9.954}{2.02} = 4.93 \tag{5.5}$$

From Equation (5.4), the theoretical gain is given as

$$\frac{V_O}{V_{IN}} = 1 + \frac{R_2}{R_1} = 1 + \frac{400\text{ K}}{100\text{ K}} = 5 \tag{5.6}$$

From Equations 5.5 and 5.6, it can be seen that the measured gain is very close to the theoretically calculated gain.

5.4 Weighted Summer

A circuit for the weighted summer is shown in Figure 5.14. Using Ohm's Law, we get

$$I_1 = \frac{V_1}{R_1}, I_2 = \frac{V_2}{R_2}, \ldots, I_n = \frac{V_n}{R_n} \tag{5.7}$$

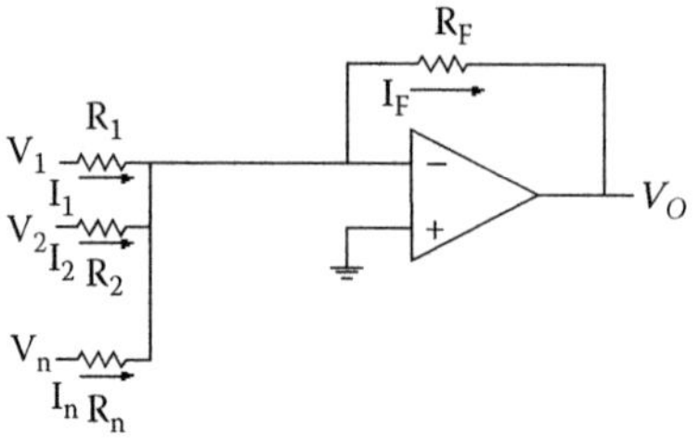

FIGURE 5.14
Weighted summer circuit.

Using Kirchhoff's Current Law, we get

$$I_F = I_1 + I_2 + \cdots I_N \tag{5.8}$$

$$V_O = -I_F R_F \tag{5.9}$$

Substituting Equations 5.7 and 5.8 into Equation 5.9, we have

$$V_O = -\left(\frac{R_F}{R_1}V_1 + \frac{R_F}{R_2}V_2 + \cdots \frac{R_F}{R_N}V_N\right) \tag{5.10}$$

The weighted summer has the following applications. It is used to add dc offset to an alternating current signal, as in a LED modulation circuit, where the dc component is needed to keep the LED in its linear operating range. It is also used to sum several signals in audio mixers. In the Example 5.3, the weighted summer is used to add direct current to an alternating current signal.

Example 5.3 Summer Amplifier

Use the Analog Discovery board to realize the expression

$$V_O(t) = -(200 + 50\sin(400\pi t))\ \text{mV} \tag{5.11}$$

Solution:
Figure 5.15 shows a circuit that can be used to realize Equation 5.11. Note that R1 = R2 = R3. From Equation 5.10, we get

$$V_o(t) = -(V_1(t) + V_2(t))$$

where

$$V_1(t) = 50\sin(400\pi t)\ \text{mV}$$

$$V_2(t) = 200\ \text{mV}$$

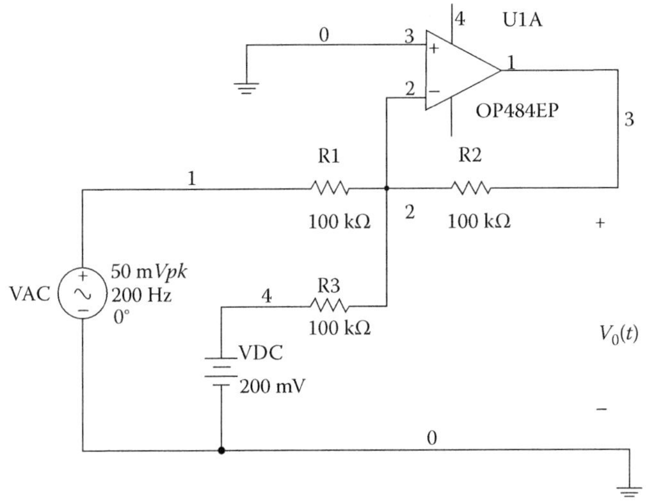

FIGURE 5.15
Op amp summer circuit for Equation 5.11.

Build the Circuit:
Use a breadboard to build the circuit shown in Figure 5.15. Connect the AWG1 (yellow wire) to node 1 of the circuit. Also, connect the scope channel 1 positive (orange wire) to node 1. In addition, connect the scope channel 1 negative (orange-white wire) to node 0. Furthermore, connect the ground (black wire) to node 0 of the circuit. Moreover, connect the scope channel 2 positive (blue wire) to node 3, and the scope channel 2 negative (blue-white wire) to node 0.

For the op amp, connect the inverting input (pin #2) to node 2 of the circuit. In addition, connect the non-inverting input (pin #3) to node 0. Moreover, connect the op amp output (pin #1) to node 3. Furthermore, connect the op amp pin #4 to the 5 V voltage source (red wire of the Analog Discovery). Moreover, connect the op amp pin #11 to the −5 V voltage source (white wire of the Analog Discovery).

Generation of AC Sine Wave:
Click on the WaveGen in the main WaveForms screen. Go to Basic → Sine → Frequency (set to 200 Hz) → Amplitude (set to 50 mV) → Offset (set to 0 V) → Symmetry (set to 50%) → Click Run AWG1.

Generation of the DC Input Voltage:
Select Channels → Channel 2 (AWG2) → Go to Basic → dc → offset (set to 200 mV) → Click Run AWG2. Figure 5.16 shows the Analog Discovery WaveGen setup for AWG1 and AWG2.

Activation of the Power Supply:
Click Voltage in the main WaveForms screen → Click Power to turn on the voltage sources → Click V+ to turn on the 5 V source → Click V− to turn on the −5 V voltage source → Upon completion of the measurements, click Power again to turn it off.

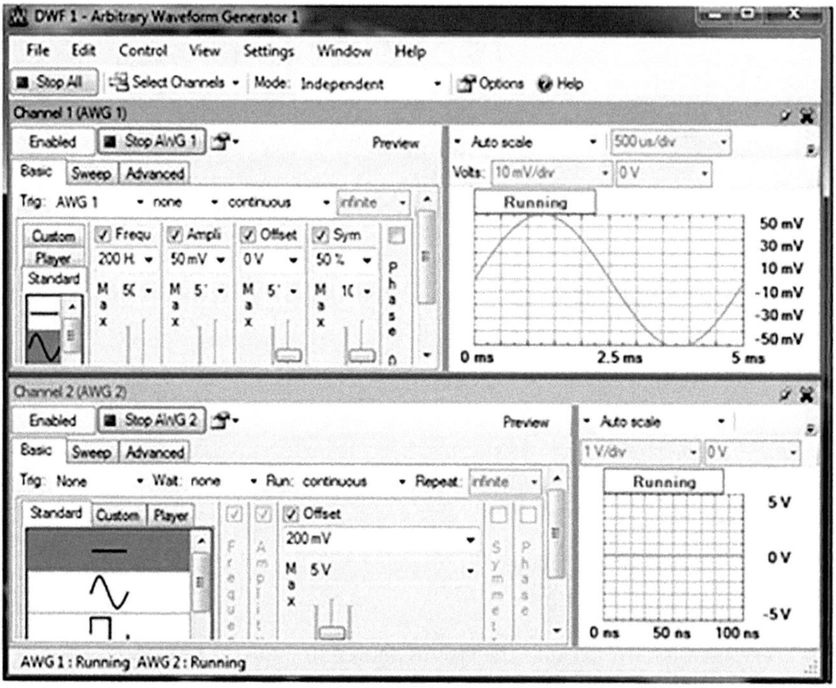

FIGURE 5.16
Sinusoidal waveform and dc signal produced by AWG1 and AWG2, respectively.

Activation of Scope:
Click on the Scope in the main WaveForms screen. Go to Run → Autoset → Measure → Add → Channel 1 → Horizontal → Frequency → Add Selected Measurement → Add → Channel 1 → Vertical → Peak to Peak → Add Selected Instrument → Add → Channel 2 → Vertical → Peak to Peak → Add → Channel 2 → Vertical → Average → Add Selected Measurement → Close Add Instrument. Figure 5.17 shows the input and output waveforms displayed on the scope.

From Figure 5.17, it can be seen that the input and output waveforms are out of phase by 180°. In addition, the output voltage has a dc component of about 212.14 mV.

5.5 Integrator and Differentiator

A basic integrator circuit is shown in Figure 5.18. In the time domain

$$\frac{V_{in}}{R_1} = I_R \text{ and } I_C = -C\frac{dV_O}{dt} \tag{5.12}$$

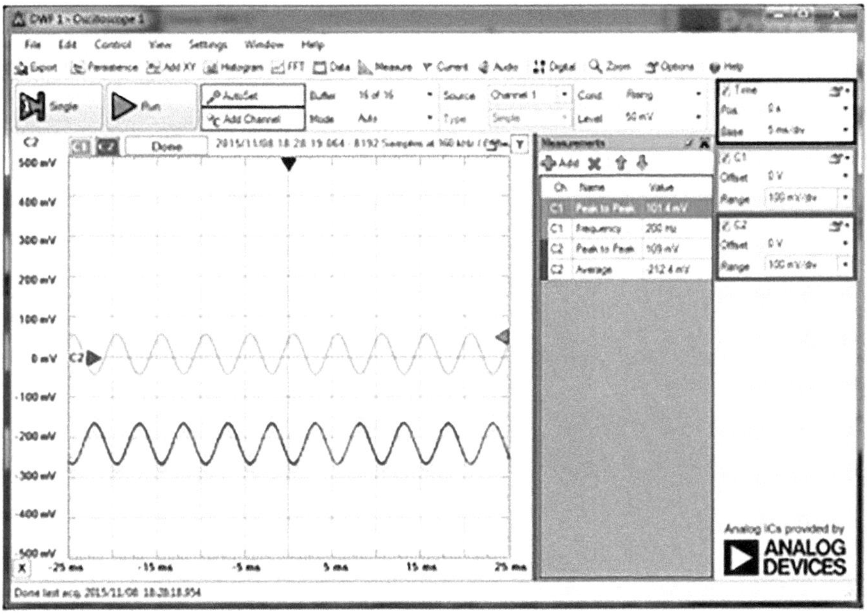

FIGURE 5.17
Display of input sinusoid (scope channel 1) and output (scope channel 2) waveforms of the summer amplifier.

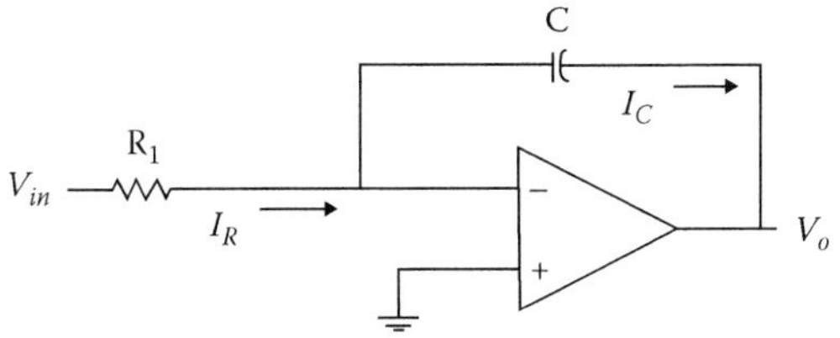

FIGURE 5.18
Op amp inverting integrator.

Since $I_R = I_C$

$$V_O(t) = -\frac{1}{R_1 C}\int_0^t V_{in}(t)\, d\tau + V_O(0) \tag{5.13}$$

The above circuit is termed the Miller integrator. The integrating time constant is CR_1. It behaves as a lowpass filter, which passes low frequencies and attenuates high frequencies. However, at dc the capacitor becomes open circuited, and there is no longer a negative feedback from the output to the input. The output voltage then saturates. To provide finite closed-loop gain at dc, a resistance R_2 is connected in parallel with the capacitor. The circuit is shown in Figure 5.19. The resistance R_2 is chosen such that R_2 is far greater than R_1.

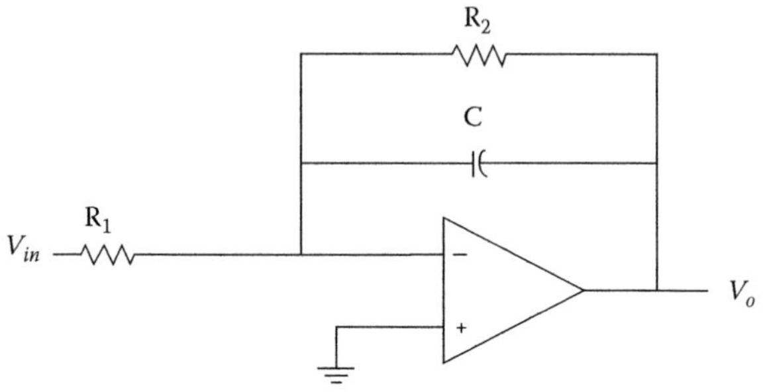

FIGURE 5.19
Miller integrator with finite closed-loop gain at dc.

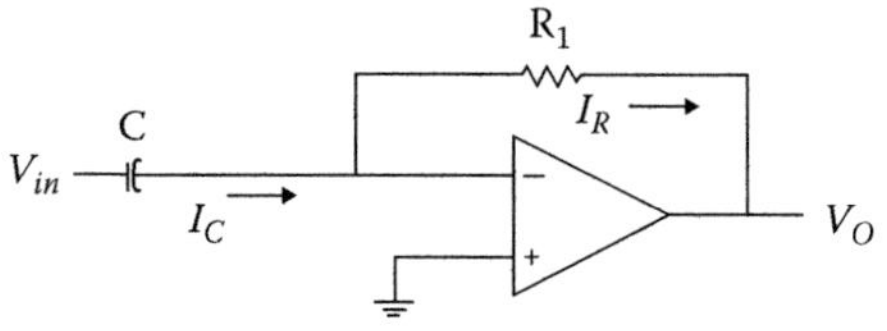

FIGURE 5.20
Op amp differentiator circuit.

Figure 5.20 is an op amp differentiator circuit. In the time domain

$$I_C = C\frac{dV_{in}}{dt}, \text{ and } V_O(t) = -I_R R_1 \tag{5.14}$$

Since

$$I_C(t) = I_R(t)$$

we have

$$V_O(t) = -CR_1 \frac{dV_{in}(t)}{dt} \tag{5.15}$$

Differentiator circuits differentiate input signals. This implies that if an input signal is rapidly changing with respect to time, the output of the differentiator circuit will appear "spike-like." Example 5.4 shows how to use the integrator to convert a square wave to a triangular wave.

Example 5.4 Converting Square Wave to Triangular Waveform

For Figure 5.21, use the Analog Discovery board to display the output voltage if the input is a square wave with the following parameters: frequency is 1 kHz, and amplitude is 2 V peak-to-peak with zero dc offset voltage.

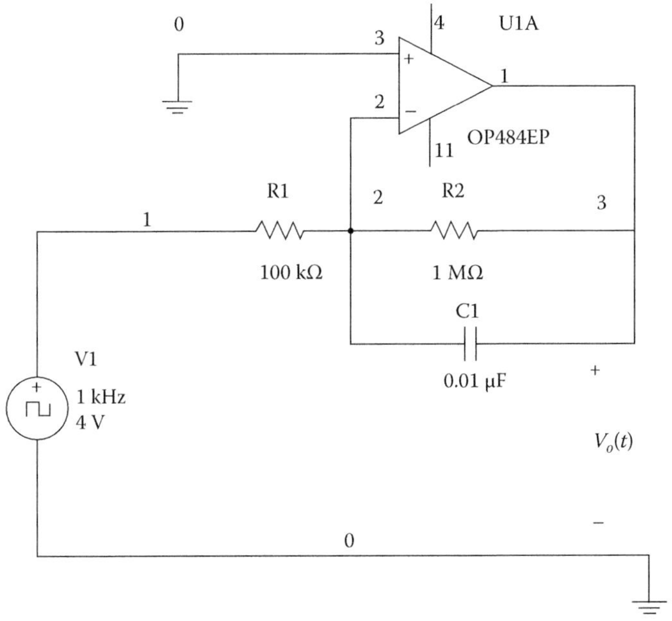

FIGURE 5.21
Op amp integrator circuit for converting square wave to triangular waveform.

Solution:

Build the Circuit:

Use a breadboard to build the circuit shown in Figure 5.21. Connect the AWG1 (yellow wire) to node 1 of the circuit. Also, connect the scope channel 1 positive (orange wire) to node 1. In addition, connect the scope channel 1 negative (orange-white wire) to node 0. Furthermore, connect the ground (black wire) to node 0 of the circuit. Moreover, connect the scope channel 2 positive (blue wire) to node 3, and the scope channel 2 negative (blue-white wire) to node 0.

For the op amp, connect the inverting input (pin #2) to node 2 of the circuit. In addition, connect the non-inverting input (pin #3) to node 0. Moreover, connect the op amp output (pin #1) to node 3 of the circuit. Furthermore, connect the op amp pin #4 to the 5 V voltage source (red wire of the Analog Discovery). Furthermore, connect the op amp pin #11 to the −5 V voltage source (white wire of the Analog Discovery).

Generation of Square Wave:

Click on the WaveGen in the main WaveForms screen. Go to Basic → Square Wave → Frequency (set to 1000 Hz) → Amplitude (set to 2 V) → Offset (set to 0 V) → Symmetry (set to 50%) → Click Run AWG1. Figure 5.22 shows the Analog Discovery WaveGen setup for generating the square wave.

Activation of the Power Supply:

Click Voltage in the main WaveForms screen → Click Power to turn on the voltage sources → Click V+ to turn on the 5 V source → Click V− to

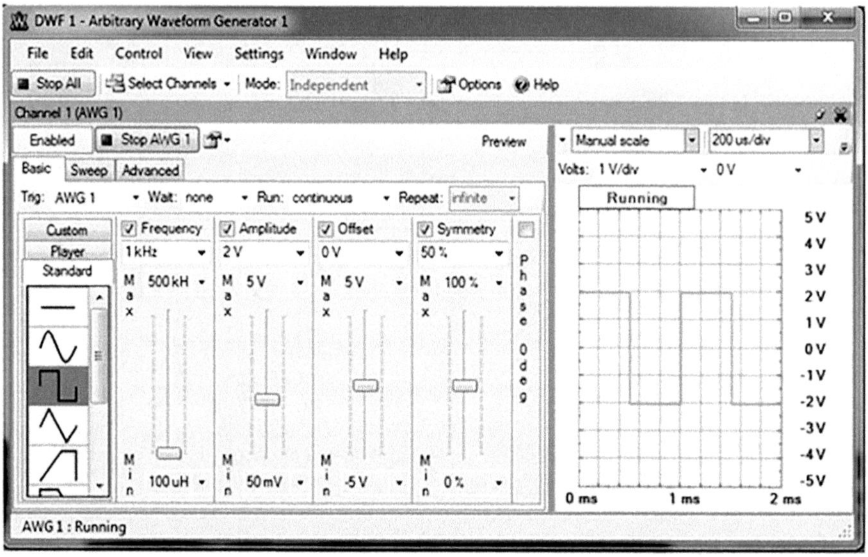

FIGURE 5.22
Square waveform produced by AWG1.

turn on the –5 V voltage source → Upon completion of the measurements, click Power again to turn it off.

Activation of Scope:
Click on the Scope in the main WaveForms screen. Go to Run → Autoset → Measure → Add → Channel 1 → Horizontal → Frequency → Add Selected Measurement → Add → Channel 1 → Vertical → Peak to Peak → Add Selected Instrument → Add → Channel 2 → Vertical → Peak to Peak → Add Selected Measurement → Close Add Instrument. Figure 5.23 shows the input and output waveforms displayed on the scope.

It can be seen from Figure 5.23, that whereas the input signal is a square wave, the output is a triangular waveform. The circuit integrates the input square wave to obtain an output signal that is a triangular waveform.

5.6 Active Filter Circuits

Electronic filter circuits are circuits that can be used to attenuate particular band(s) of frequency and also pass other band(s) of frequency. The following types of filters are discussed in this section: lowpass, bandpass, highpass, and band-reject. The filters have passband, stopband, and transition band. The order of the filter determines the transition from the passband to stopband.

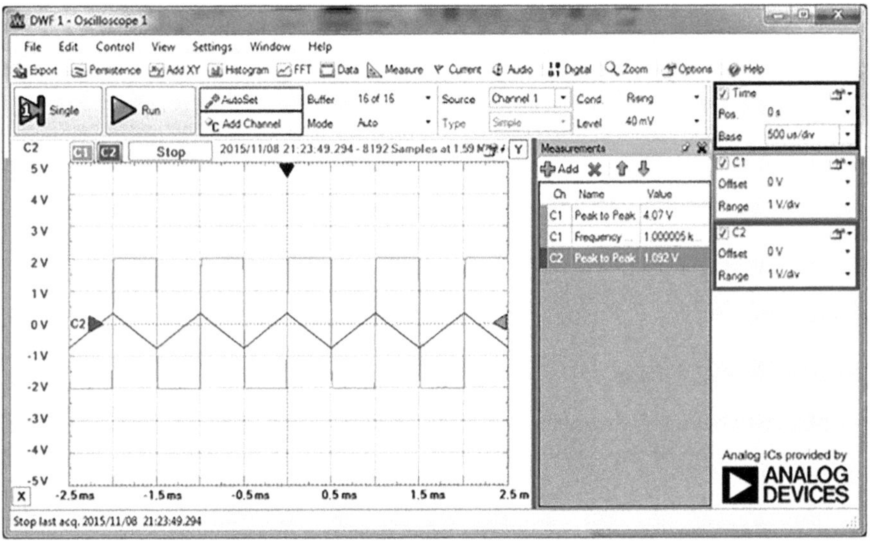

FIGURE 5.23
Display of input square (scope channel 1) and output triangular (scope channel 2) waveforms of the integrator circuit.

5.6.1 Lowpass Filters

Lowpass filters pass low frequencies and attenuate high frequencies. The transfer function of a first order lowpass filter has the general form

$$H(s) = \frac{k}{s + \omega_0} \tag{5.16}$$

A circuit that can be used to implement a first-order lowpass filter is shown in Figure 5.24.

The voltage transfer function for Figure 5.24 is

$$H(s) = \frac{V_0(s)}{V_{in}(s)} = \frac{k}{1 + sR_1C_1} \tag{5.17}$$

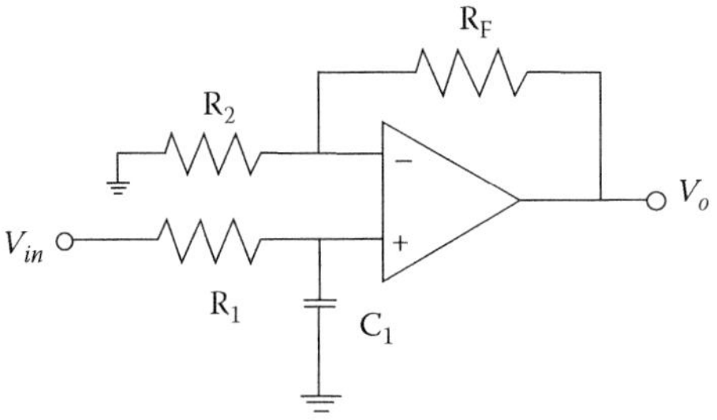

FIGURE 5.24
First-order lowpass filter.

with dc gain, k, given as

$$k = 1 + \frac{R_F}{R_2} \tag{5.18}$$

and the cut-off frequency, f_0, is

$$f_0 = \frac{1}{2\pi R_1 C_1} \tag{5.19}$$

The first-order filter exhibits –20 dB/decade roll-off in the stopband.

5.6.2 Highpass Filters

Highpass filters pass high frequencies and attenuate low frequencies. The transfer function of the first-order highpass filter has the general form

$$H(s) = \frac{ks}{s + \omega_0} \tag{5.20}$$

The circuit, shown in Figure 5.25, can be used to implement the first-order highpass filter. It is basically the same as Figure 5.24, except that the positions of R_1 and C_1 in Figure 5.24 have been interchanged.

For Figure 5.25, the voltage transfer function is

$$H(s) = \frac{V_O}{V_{in}}(s) = \frac{s}{s + \dfrac{1}{R_1 C_1}}\left(1 + \frac{R_F}{R_2}\right) \tag{5.21}$$

where

$$k = \left(1 + \frac{R_F}{R_2}\right) \tag{5.22}$$

k = gain at very high frequency

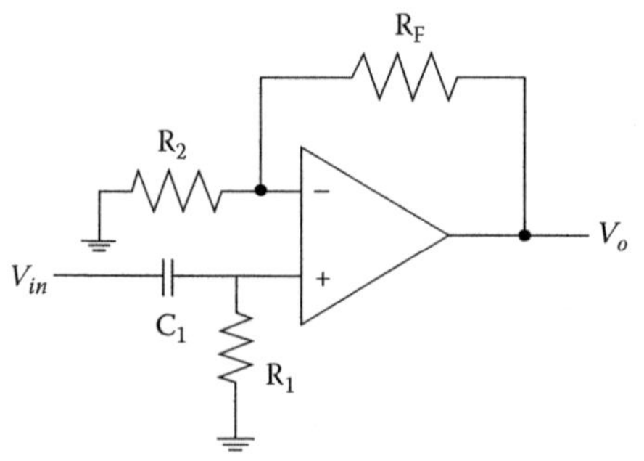

FIGURE 5.25
First-order highpass filter.

and the cut-off frequency at 3dB gain is

$$f_0 = \frac{1}{2\pi R_1 C_1} \tag{5.23}$$

Although the filter, shown in Figure 5.25, passes all signal frequencies higher than f_0, the high-frequency characteristic is limited by the bandwidth of the op amp.

5.6.3 Bandpass Filters

Bandpass filters pass a band of frequencies and attenuate other bands. The filter has two cut-off frequencies f_L and f_H. We assume that $f_H > f_L$. All signal frequencies lower than f_L or greater than f_H are attenuated. The general form of the transfer function of a bandpass filter is

$$H(s) = \frac{k\left(\frac{\omega_c}{Q}\right)s}{s^2 + \left(\frac{\omega_c}{Q}\right)s + \omega_c^2} \tag{5.24}$$

where

k is the passband gain.

ω_c is the center frequency in rad/s.

The quality factor Q is related to the 3-dB bandwidth and the center frequency by the expression

$$Q = \frac{\omega_c}{BW} = \frac{f_c}{f_H - f_L} \tag{5.25}$$

Bandpass filters with $Q < 10$ are classified as wide bandpass. On the other hand, bandpass filters with $Q > 10$ are considered narrow bandpass. Wideband pass filters may be implemented by cascading lowpass and highpass filters. The order of the bandpass filters is the sum of the highpass and lowpass sections. The advantages of this arrangement are that the fall-off, bandwidth and midband gain can be set independently.

Figure 5.26 shows the wideband pass filter, built using first-order highpass and first-order lowpass filters. The magnitude of the voltage gain is the product of the voltage gains of both the highpass and lowpass filters.

A narrowband pass filter normally has a high Q-value. A circuit that can be used to implement a narrowband filter is a multiple feedback filter, as shown in Figure 5.27.

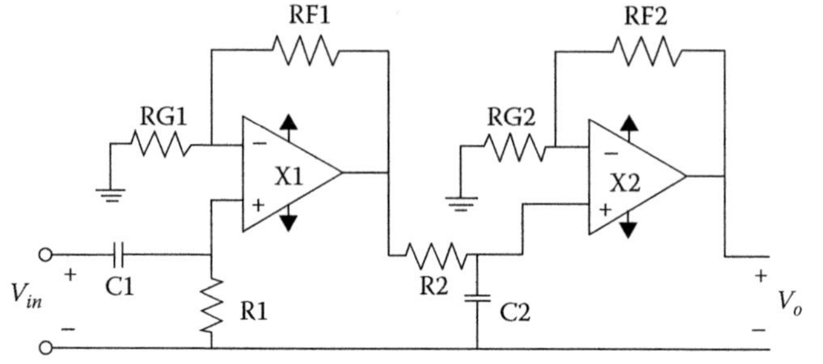

FIGURE 5.26
Second-order wideband pass filter.

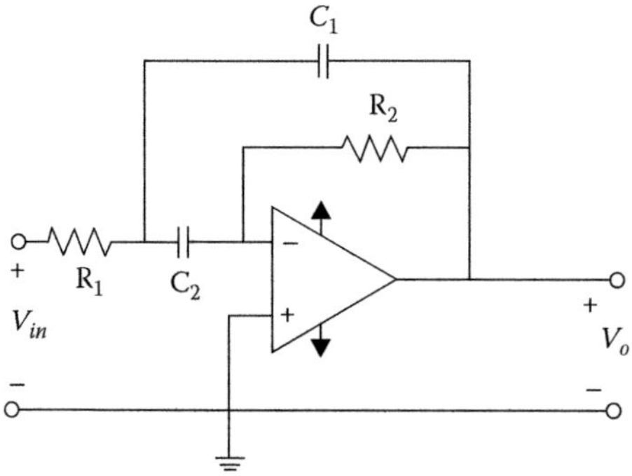

FIGURE 5.27
Multiple-feedback bandpass filter.

5.6.4 Band-Reject Filters

A band-reject is used to eliminate a specific band of frequencies. It is normally used in communication and biomedical instruments to eliminate unwanted frequencies. The general form of the transfer function of the band-reject filter is

$$H_{BR} = \frac{k_{PB}\left(s^2 + \omega_C^2\right)}{s^2 + \left(\frac{\omega_C}{Q}\right)s + \omega_C^2} \tag{5.26}$$

where

k_{PB} is the passband gain.

ω_C is the center frequency of the band-reject filter.

Band-reject filters are classified as wideband reject ($Q < 10$) and narrowband reject filters ($Q > 10$). Narrowband reject filters are commonly called notch filters. The wideband reject filter can be implemented by summing the responses of the highpass section and the lowpass section through a summing amplifier. The block diagram arrangement is shown in Figure 5.28.

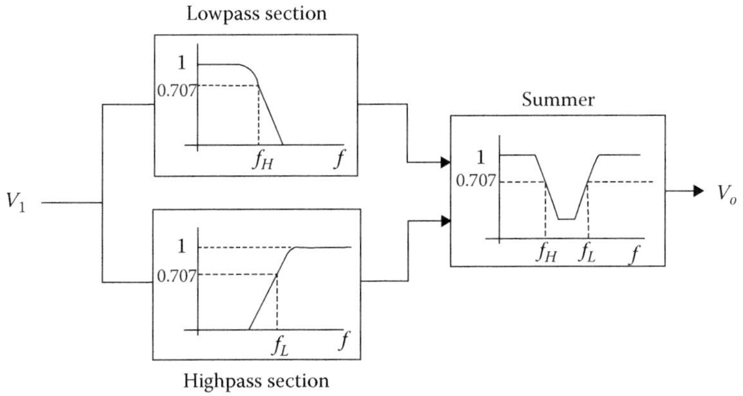

FIGURE 5.28
Block diagram of wideband reject filter.

The order of the band-reject filter is dependent on the order of lowpass and highpass sections. There are two important requirements for implementing the wideband reject filter using the scheme shown in Figure 5.28:

1. The cut-off frequency, f_L, of the highpass filter section must be greater than the cut-off frequency, f_H, of the lowpass filter section.
2. The passband gains of the lowpass and highpass sections must be equal.

Narrowband reject filters or notch filters can be implemented by using a Twin-T network. This circuit consists of two parallel T-shaped networks, as shown in Figure 5.29.

Normally, R1=R2=R, R3=R/2, C1=C2=C and C3=2C. The R1–C3–R2 section is a lowpass filter with corner frequency $f_C = (4\pi RC)^{-1}$. The C1–R3–C2 is a highpass filter with corner frequency $f_C = (\pi RC)^{-1}$.

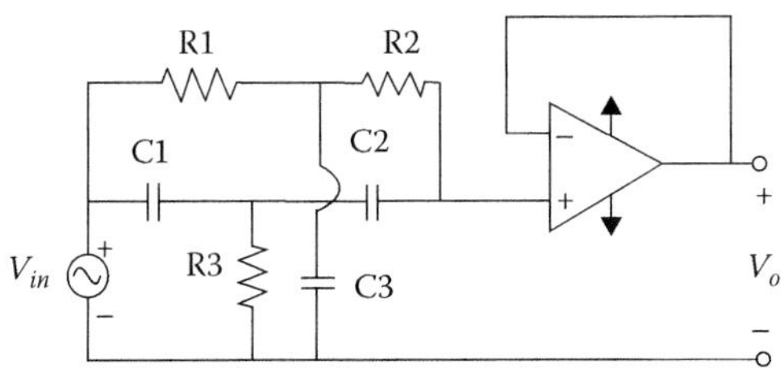

FIGURE 5.29
Narrowband reject Twin-T network.

The center frequency or the notch frequency is

$$f_C = \frac{1}{2\pi RC} \tag{5.27}$$

At the notch frequency given by Equation 5.27, the phases of the two filters cancel out. Example 5.5 explores the characteristics of the notch filter through the use of the Analog Discovery board.

Example 5.5 Frequency Response of a Notch Filter

Figure 5.30 shows a notched filter built using an op amp. Determine the center frequency and the bandwidth of the filter.

Solution:

Build the Circuit:

Use a breadboard to build the circuit shown in Figure 5.30. Connect the AWG1 (yellow wire) to node 2 of the circuit. Also, connect the scope channel 1 positive (orange wire) to node 2. In addition, connect the scope channel 1 negative (orange-white wire) to node 0. Moreover, connect the ground (black wire) to node 0 of the circuit. Furthermore, connect the scope channel 2 positive (blue wire) to node 4, and the scope channel 2 negative (blue-white wire) to node 0.

For the op amp, connect the inverting input (pin 2) to node 4 of the circuit. In addition, connect the non-inverting input (pin #3) to node 5. Moreover, connect the op amp output (pin #1) to node 3. Furthermore, connect the op amp pin #4 to the 5 V voltage source (red wire of the ADM). Furthermore, connect the op amp pin #11 to −5 V voltage source (white wire of the ADM).

Generation of AC Sine Wave:

Click on the WaveGen in the main WaveForms screen. Go to Basic → Sine → Frequency (set to 1000 Hz) → Amplitude (set to 1.0 V) → Offset (set to 0 V)

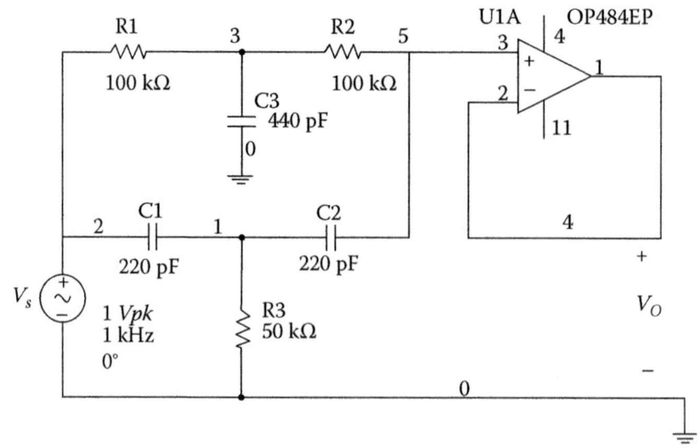

FIGURE 5.30
Band-reject filter.

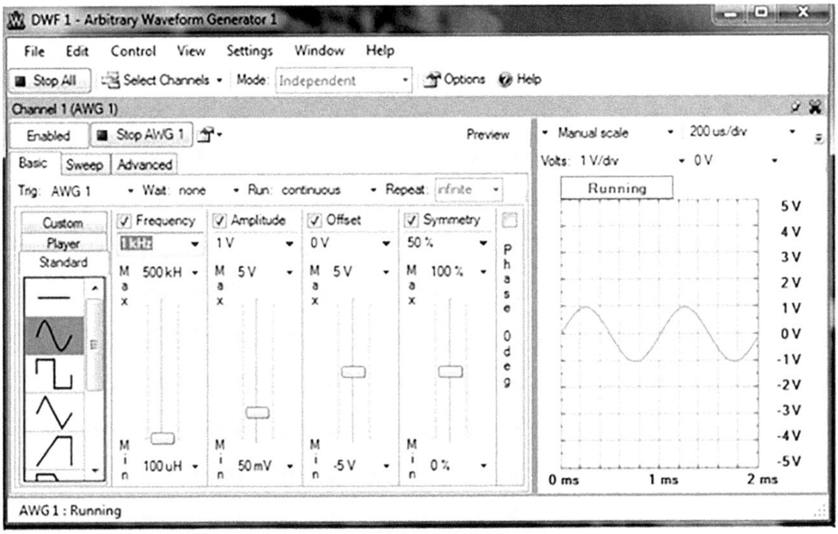

FIGURE 5.31
Analog Discovery WaveGen setup for obtaining frequency response.

→ Symmetry (set to 50%) → Click Run AWG1. Figure 5.31 shows the Analog Discovery WaveGen setup for generating a sinusoidal waveform.

Activation of the Power Supply:
Click Voltage in the main WaveForms screen → Click Power to turn on the voltage sources → Click V+ to turn on the 5 V source → Click V– to turn on the –5 V voltage source → Upon completion of the measurements, click Power again to turn it off.

Activation of Network Analyzer:
The Network Analyzer uses both oscilloscope channels as the input and output channels. Channel 1 is for measuring the input signal and Channel 2 is for measuring the output signal.

Open up the Network Analyzer that can be found under the More Instruments tab. Use the following values for setting up the Network Analyzer:

Start Frequency: 100 Hz
Stop Frequency: 100 kHz
Offset: 0 V
Input Signal Amplitude: 1 V
Max Filter Gain: 1X
Bode Scale: Magnitude—Top 5 dB, Range of 40 dB
Phase—Top 90°, Range of 180°
Scope Channel Gain: Channel 1: 1X; Channel 2: 1X
Obtain a single sweep in frequency by clicking Single. This provides a Bode plot representation of the frequency response of the circuit. The Bode plot obtained is shown in Figure 5.32.

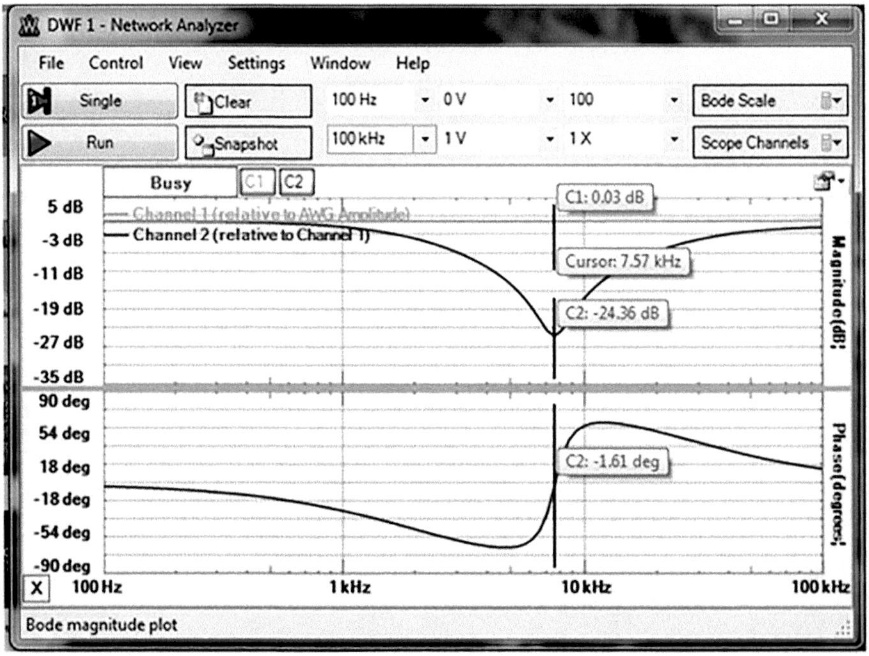

FIGURE 5.32
Magnitude and phase response of the RLC circuit of Figure 5.30.

From Figure 5.32, the following measurements were obtained:

Center frequency is 7.57 kHz.
Low cut-off frequency is 1.78 kHz.
High cut-off frequency is 36.66 kHz.
Bandwidth = 36.66 – 1.78 kHz = 34.88 kHz.

The theoretical value of the center frequency of Figure 5.30, given by Equation (5.27), is

$$f_C = \frac{1}{2\pi RC} = \frac{1}{2\pi(10^5)(220)(10^{-12})} = 7.23\text{ KHz} \tag{5.28}$$

The calculated value of the center frequency (i.e. 7.23 kHz) is close to the measured value of 7.57 kHz.

PROBLEMS

Problem 5.1 For the circuit shown in Figure P5.1, (1) build the circuit and use the Analog Discovery board to find the gain. (2) Compare the gain obtained in part (1) to the gain calculated using Equation 5.1.

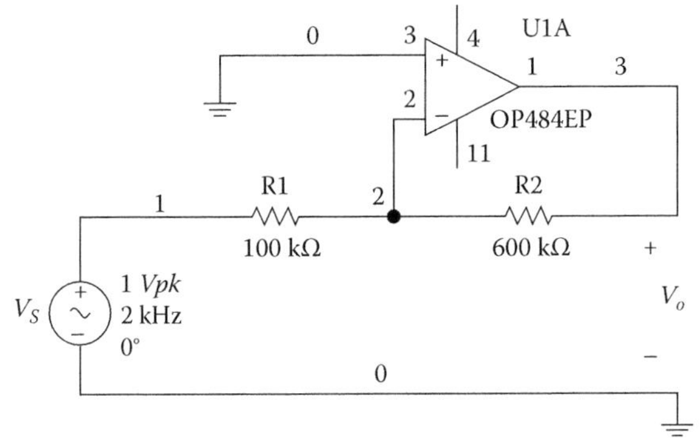

FIGURE P5.1
Inverting amplifier.

Problem 5.2 For the circuit shown in Figure P5.2, (1) build the circuit and use the Analog Discovery board to find the gain. (2) Compare the gain obtained in part (1) to the gain calculated using Equation 5.1.

Problem 5.3 The circuit shown in Figure P5.3 can be used to add the first two terms of the Fourier series expansion of a square wave. Build the circuit and use the Analog Discovery board to display the output waveform. How close is the output waveform compared to a square wave of a fundamental frequency of 2 kHz?

Problem 5.4 For Figure P5.4, (1) what will be the output waveform if the frequency of the input square waveform is 2 kHz? (2) Determine the peak-to-peak value of the output waveform when the frequency of the input waveform is (i) 4 kHz, (ii) 6 kHz, and (iii) 8 kHz. (3) Why does the peak-to-peak voltage of the output waveform decrease with increase in the input frequency?

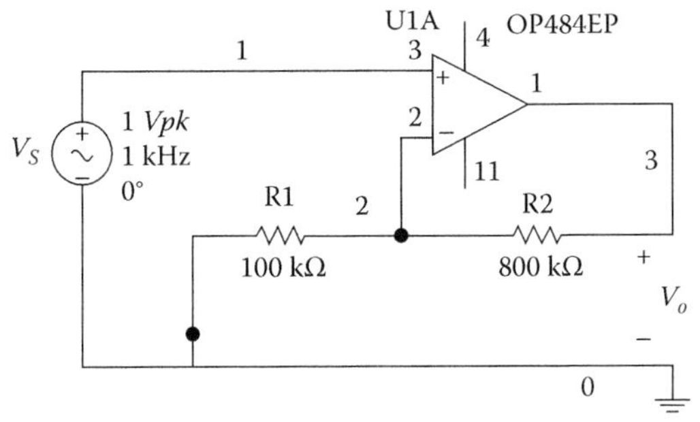

FIGURE P5.2
Non-inverting amplifier.

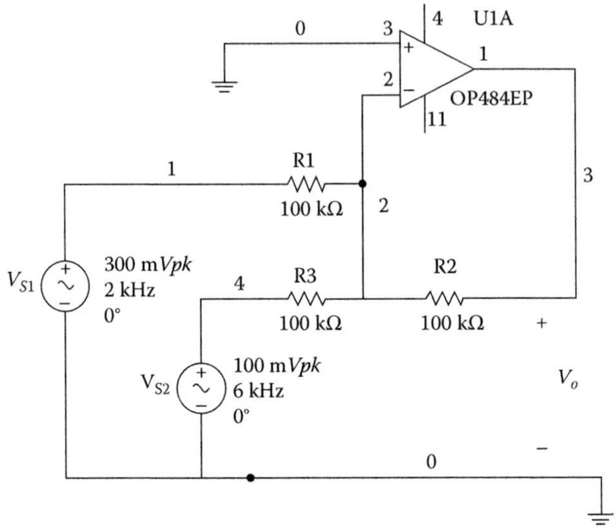

FIGURE P5.3
Summer amplifier for summing two terms of Fourier series terms.

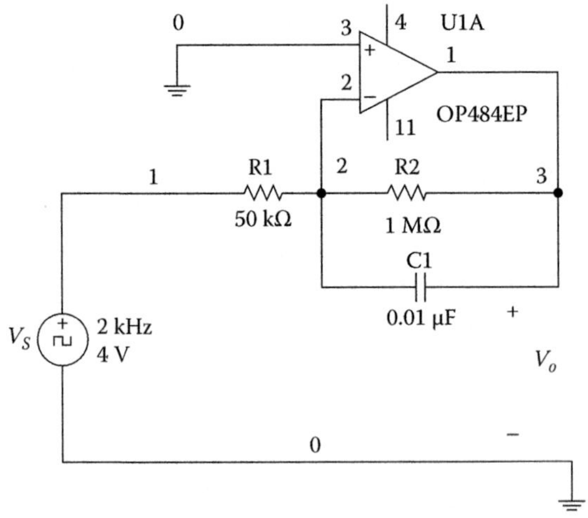

FIGURE P5.4
Integrator circuit.

Problem 5.5 For the circuit shown in Figure 5.21, if the frequency of the square wave is decreased to (1) 500 Hz, and (2) 50 Hz, determine the output voltage. Why is the output voltage of the integrator no longer a triangular waveform at a frequency of 50 Hz?

Problem 5.6 Build the circuit shown in Figure P5.6. If the input is a triangular waveform with frequency of 1 kHz, capture the output voltage by the use

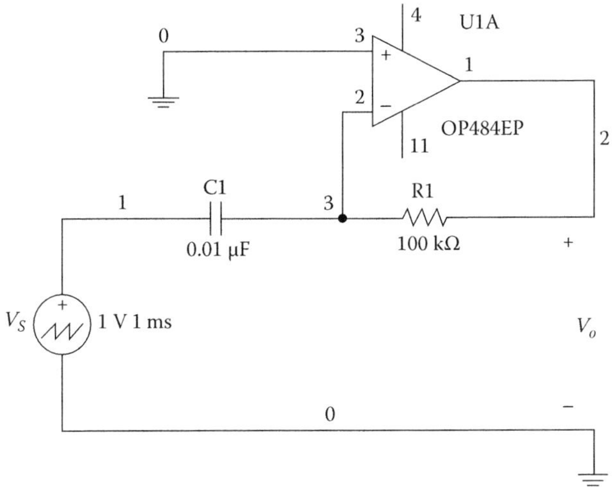

FIGURE P5.6
Differentiator circuit.

of the Analog Discovery board. Does the output indicate differentiation of the input signal?

Problem 5.7 Build the circuit shown in Figure P5.7. Use the Network Analyzer of the Analog Discovery board to: (1) Obtain the Bode plot of the filter. (2) Measure the cut-off frequency. (3) Obtain the low-frequency gain of the circuit. Compare the measured values, obtained in parts (2) and (3), with those calculated by using Equations 5.18 and 5.19.

Problem 5.8 Build the circuit shown in Figure P5.8. Use the Network Analyzer of the Analog Discovery board to: (1) Obtain the Bode plot of the filter. (2) Measure the cut-off frequency. (3) Obtain the low-frequency gain of the circuit.

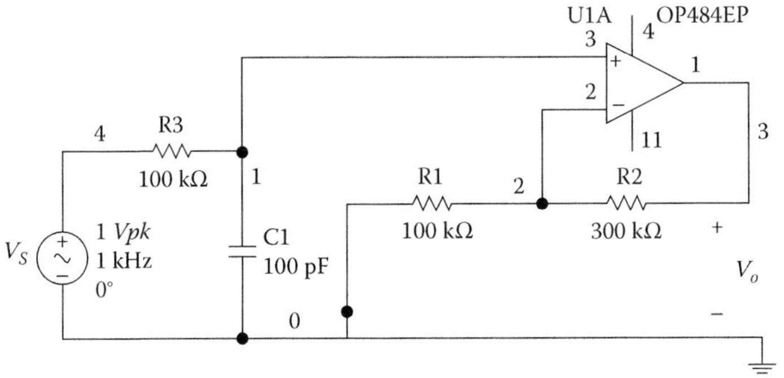

FIGURE P5.7
Op amp lowpass filter.

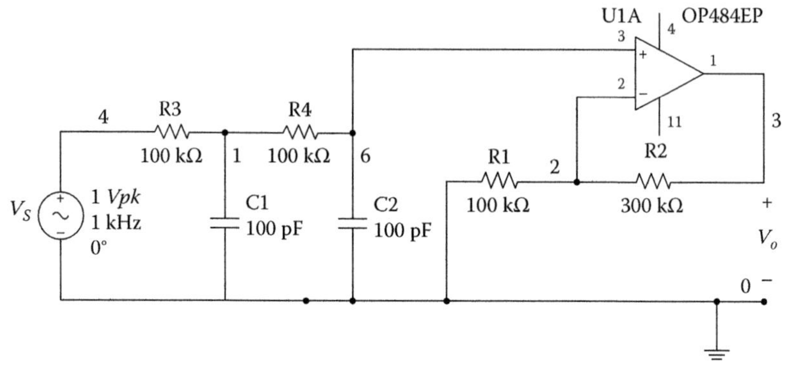

FIGURE P5.8
Op amp two-stage lowpass filter.

Problem 5.9 Build the circuit shown in Figure P5.9. Use the Network Analyzer of the Analog Discovery board to: (1) Obtain the Bode plot of the filter. (2) Measure the cut-off frequency. (3) Obtain the high-frequency gain of the circuit. Compare the measured values, obtained in parts (2) and (3), with those calculated by using Equations 5.22 and 5.23.

Problem 5.10 Build the circuit shown in Figure P5.10. Use the Network Analyzer of the Analog Discovery board to: (1) Obtain the Bode plot of the filter. (2) Measure the cut-off frequency. (3) Obtain the high-frequency gain of the circuit.

Problem 5.11 Build the circuit shown in Figure P5.11. Use the Analog Discovery board to: (1) Measure the center frequency and (2) measure the low cut-off and high cut-off frequencies. (3) Calculate the bandwidth of the

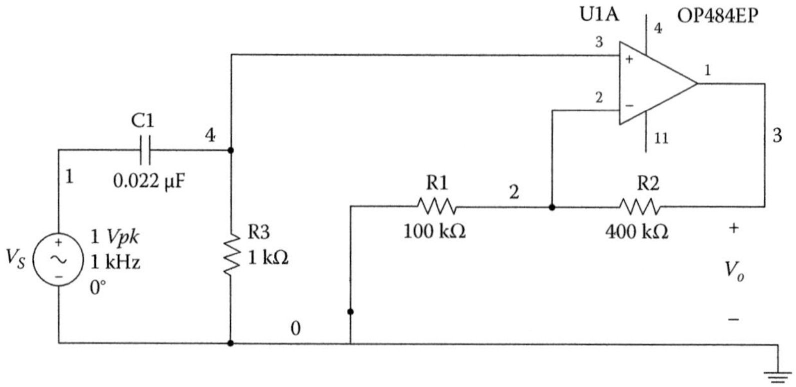

FIGURE P5.9
Op amp highpass filter.

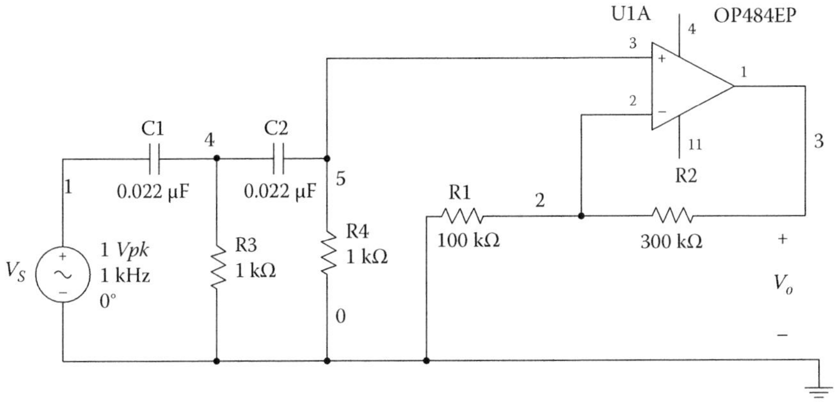

FIGURE P5.10
Op amp two-stage highpass filter.

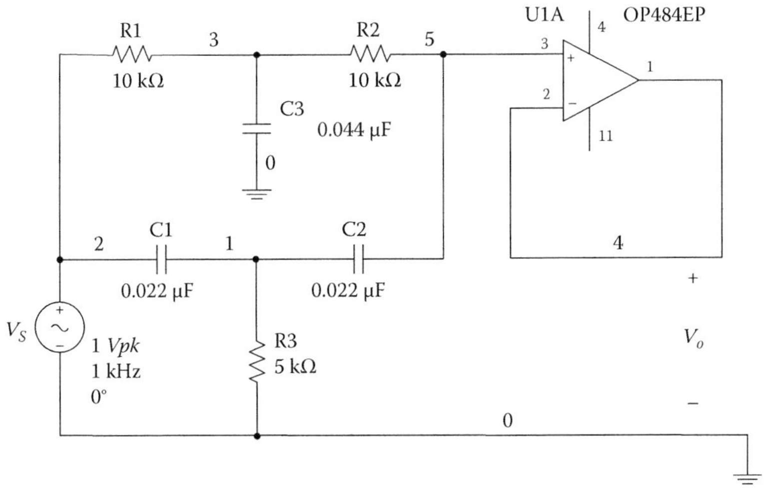

FIGURE P5.11
Op amp notched filter.

filter. (4) Compare your measured center frequency with that calculated from Equation 5.27.

Problem 5.12 For Figure P5.12, build the circuit. Use the Analog Discovery board to measure: (1) The low cut-off frequency. (2) The high cut-off frequency, and (3) the maximum gain.

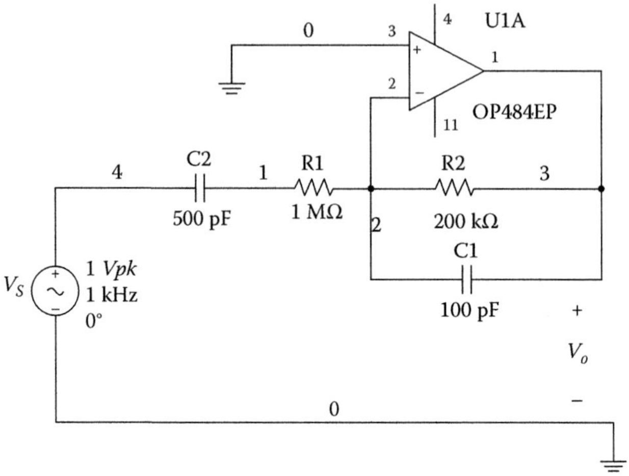

FIGURE P5.12
Op amp bandpass filter.

6

Diodes

INTRODUCTION In this chapter, the characteristics of diodes are presented. In addition, the operation of diode circuits is discussed. The Analog Discovery board is used to obtain diode I–V characteristics, and to study the characteristics of half- and full-wave rectifier circuits.

6.1 Diode Characteristics

A diode is a two-terminal device. The electronic symbol of a diode is shown in Figure 6.1a. Ideally, the diode conducts current in one direction. The current versus voltage characteristics of an ideal diode are shown in Figure 6.1b.

The I–V characteristic of a semiconductor junction diode is shown in Figure 6.2. The characteristic is divided into three regions: forward-biased, reversed-biased, and the breakdown regions.

In the forward-biased and reversed-biased regions, the current, i, and the voltage, v, of a semiconductor diode are related by the diode equation

$$i = I_S[e^{(v/nV_T)} - 1] \tag{6.1}$$

where

I_S is the reverse saturation current or leakage current,
n is an empirical constant between 1 and 2,
V_T is thermal voltage, given by

$$V_T = \frac{kT}{q} \tag{6.2}$$

and

k is Boltzmann's constant $= 1.38 \times 10^{-23}$ J/°K,
q is the electronic charge $= 1.6 \times 10^{-19}$ Coulombs,
T is the absolute temperature in °K.

At room temperature (25°C), the thermal voltage is about 25.7 mV.

6.1.1 Forward-Biased Region

In the forward-biased region, the voltage across the diode is positive. If we assume that at room temperature the voltage across the diode is greater than 0.1 V, then Equation 6.1 simplifies to

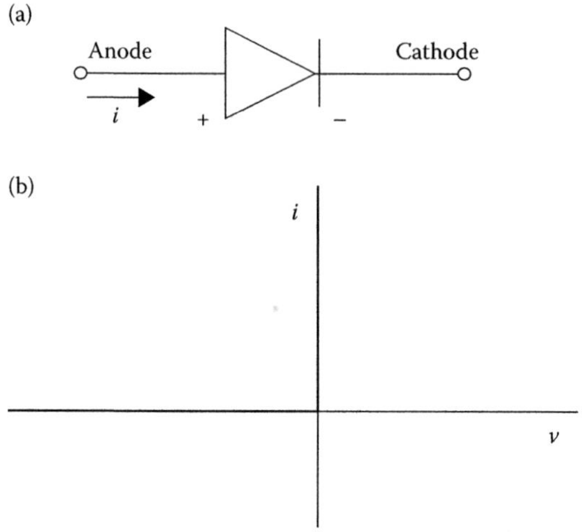

FIGURE 6.1
Ideal diode (a) electronic symbol, (b) I–V characteristics.

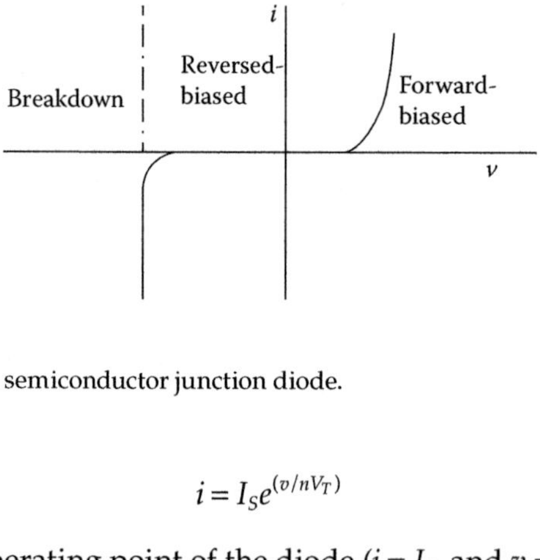

FIGURE 6.2
I–V characteristics of a semiconductor junction diode.

$$i = I_S e^{(v/nV_T)} \tag{6.3}$$

For a particular operating point of the diode ($i = I_D$ and $v = V_D$), we have

$$i_D = I_S e^{(v_D/nV_T)} \tag{6.4}$$

To obtain the dynamic resistance of the diode at a specified operating point, we differentiate Equation 6.3 with respect to v, and we have

$$\frac{di}{dv} = \frac{I_s e^{(v/nV_T)}}{nV_T}$$

$$\left.\frac{di}{dv}\right|_{v=V_D} = \frac{I_s e^{(v_D/nV_T)}}{nV_T} = \frac{I_D}{nV_T}$$

and the dynamic resistance of the diode, r_d, is

$$r_d = \frac{dv}{di}\bigg|_{v=V_D} = \frac{nV_T}{I_D} \tag{6.5}$$

From Equation 6.3, we have

$$\frac{i}{I_S} = e^{(v/nV_T)}$$

thus

$$\ln(i) = \frac{v}{nV_T} + \ln(I_S) \tag{6.6}$$

Equation 6.6 can be used to obtain the diode constants n and I_S, given the data that consists of the corresponding values of voltage and current. From Equation 6.6, a curve of v versus $\ln(i)$ will have a slope given by $1/nV_T$ and a y-intercept of $\ln(I_S)$. Example 6.1 shows how the Analog Discovery board can be used to obtain the I–V characteristics of a diode.

Example 6.1 Diode Characteristics

Use the Analog Discovery board to obtain the I–V characteristics of diode 1N4001.

Solution:
Build the Circuit:
Use a breadboard to build the circuit shown in Figure 6.3. Connect the AWG1 (yellow wire) to node 1 of the circuit. Also, connect the scope channel 1 positive (orange wire) to node 2. In addition, connect the scope channel 1 negative (orange-white wire) to node 0. Moreover, connect the ground (black wire) to node 0 of the circuit. Furthermore, connect the scope channel 2 positive (blue wire) to node 1, and the scope channel 2 negative (blue-white wire) to node 2.

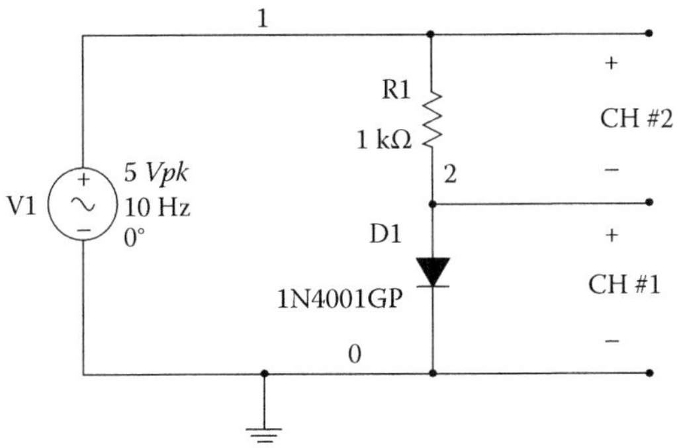

FIGURE 6.3
Diode 1N4001 circuit.

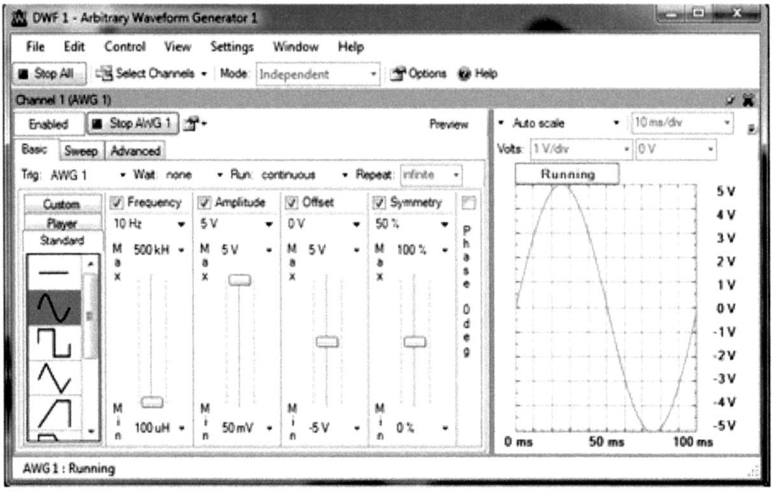

FIGURE 6.4
Analog Discovery WaveGen setup for I–V characteristic measurement.

Generation of Sine Wave:
Click on the WaveGen in the main WaveForms screen. Go to Basic → Sine → Frequency (set to 10 Hz) → Amplitude (set to 5 V) → Offset (set to 0 V) → Symmetry (set to 50%) → Click Run AWG1. Figure 6.4 shows the setup of the Analog Discovery WaveGen AWG1.

Activation of Scope:
Click on the Scope in the main WaveForms screen. Go to Run → Autoset → Measure → Add → Channel 1 → Horizontal → Frequency → Add Selected Measurement → Add → Channel 1 → Vertical → Peak to Peak → Add Selected Instrument → Add → Channel 2 → Vertical → Peak to Peak → Control → Add Mathematic Channel → Custom → Type C2/1000 (this current is calculated as the voltage across the 1000 Ω resistor divided by the resistance of 1000 Ω) → control → Add XY (this initiates the XY plot) → set X (Channel 1) and set Y (Math1) → Change Math Mode Units to A → Change the M1 range to 1 mA/V. Figure 6.5 shows the I–V characteristics of the diode.

6.2 Half-Wave Rectification

A half-wave rectifier circuit is shown in Figure 6.6. It consists of an alternating current (ac) source, a diode, and a resistor.

Assuming that the diode is ideal, the diode conducts when source voltage is positive, making

$$v_0 = v_S \quad \text{when } v_S \geq 0 \tag{6.7}$$

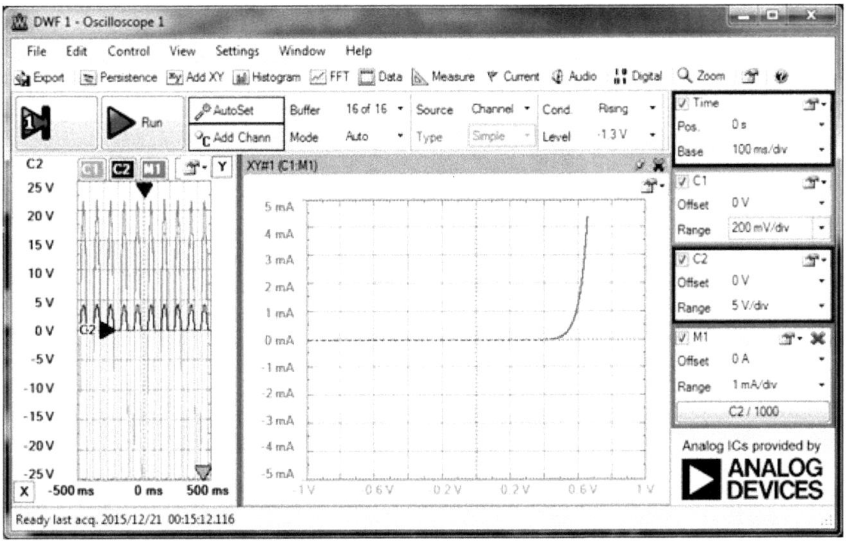

FIGURE 6.5
I–V characteristics of diode 1N4001.

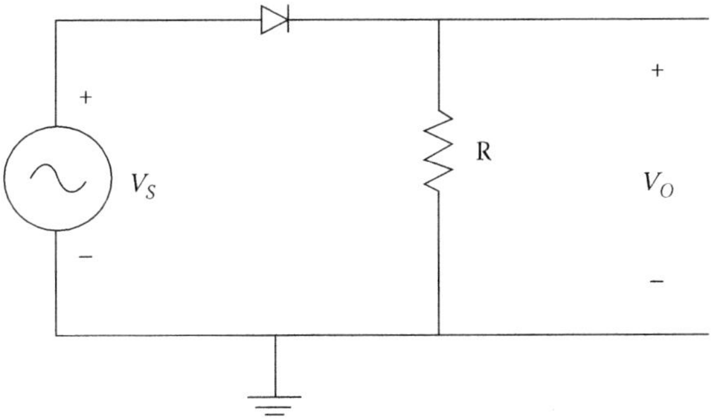

FIGURE 6.6
Half-wave rectifier circuit.

When the source voltage is negative, the diode is cut off, and the output voltage is

$$v_0 = 0 \text{ when } v_S < 0 \tag{6.8}$$

Example 6.2 shows how to use the Analog Discovery board to obtain the input and output waveforms of a half-wave rectifier.

Example 6.2 Half-Wave Rectification

Determine the voltage at the output of a half-wave rectifier circuit built using the diode 1N4001, shown in Figure 6.7.

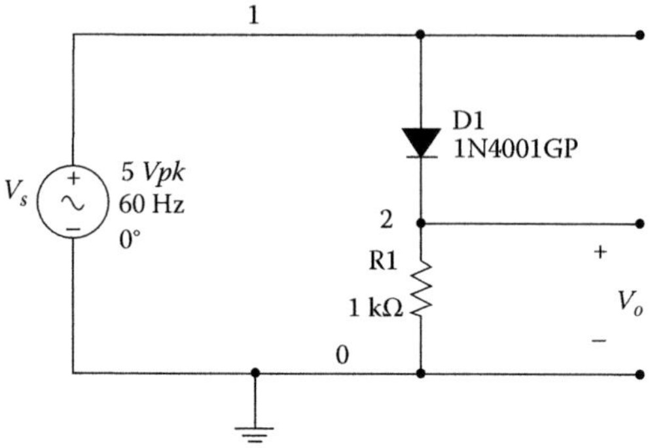

FIGURE 6.7
Half-wave rectifier circuit built using diode 1N4001.

Solution:
Build the Circuit:
Use a breadboard to build the circuit shown in Figure 6.7. Connect the AWG1 (yellow wire) to node 1 of the circuit. Also, connect the scope channel 1 positive (orange wire) to node 1. In addition, connect the scope channel 1 negative (orange-white wire) to node 0. Moreover, connect the ground (black wire) to node 0 of the circuit. Furthermore, connect the scope channel 2 positive (blue wire) to node 2, and the scope channel 2 negative (blue-white wire) to node 0.

Generation of Sine Wave:
Click on the WaveGen in the main WaveForms screen. Go to Basic → Sine → Frequency (set to 60 Hz) → Amplitude (set to 5 V) → Offset (set to 0 V) → Symmetry (set to 50%) → Click Run AWG1. Figure 6.8 shows the setup of Analog Discovery WaveGen AWG1.

Activation of Scope:
Click on the Scope in the main WaveForms screen. Go to Run → Autoset → Measure → Add → Channel 1 → Horizontal → Frequency → Add Selected Measurement → Add → Channel 1 → Vertical → Peak to Peak → Add Selected Instrument → Add → Channel 2 → Vertical → Peak to Peak → Close Add Instrument.

Figure 6.9 shows the oscilloscope display of half-wave rectification. The battery-charging circuit, which consists of a voltage source connected to a battery through a resistor and a diode, is explored in Example 6.3.

Example 6.3 Battery-Charging Circuit

A battery-charging circuit is shown in Figure 6.10. The battery voltage is 2 V. The source voltage is $v_S(t) = 5\sin(120\pi t)$ V, and the load resistance is 1000 Ω. Use the Analog Discovery board to (1) display the input voltage, (2) display the current flowing through the diode, (3) measure the conduction angle of the diode, and (4) determine the peak current.

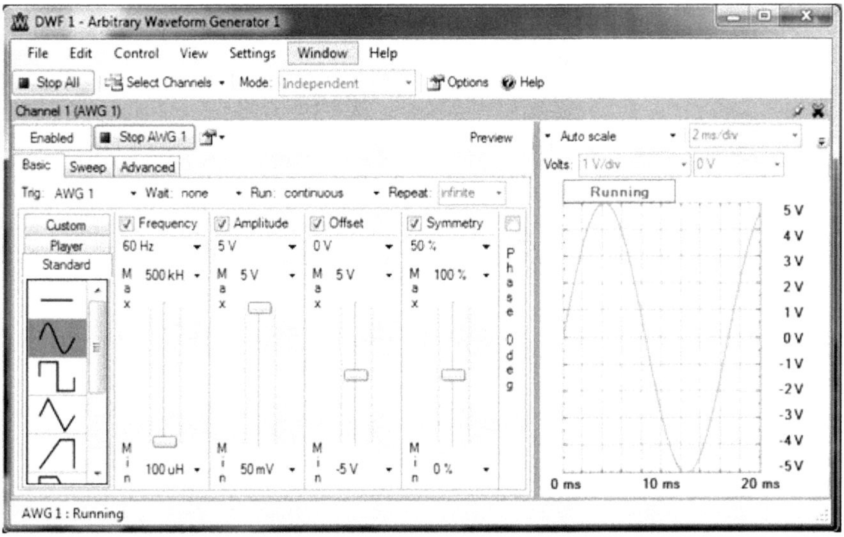

FIGURE 6.8
Analog Discovery WaveGen setup for half-wave rectification circuit.

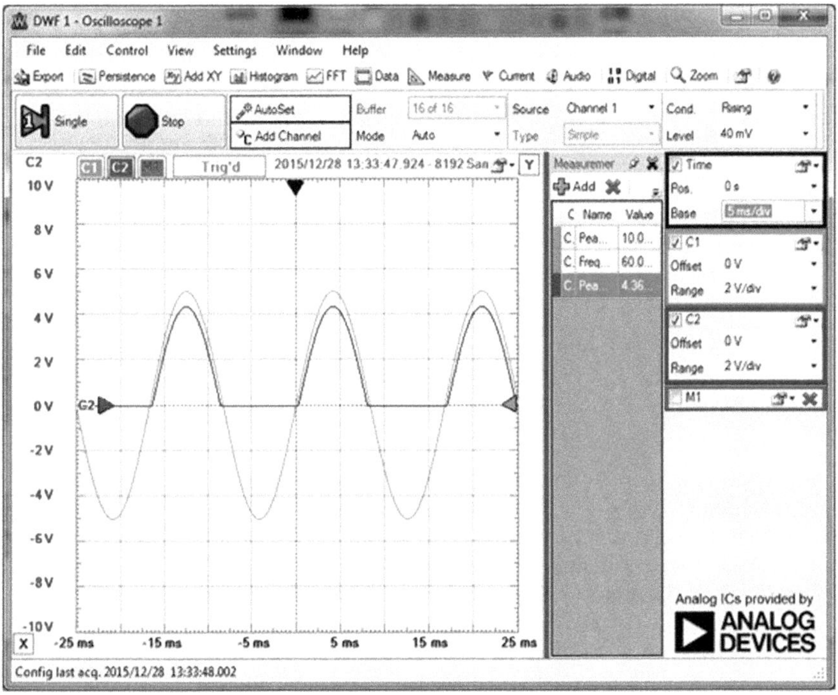

FIGURE 6.9
Half-wave rectification waveform (scope channel 2).

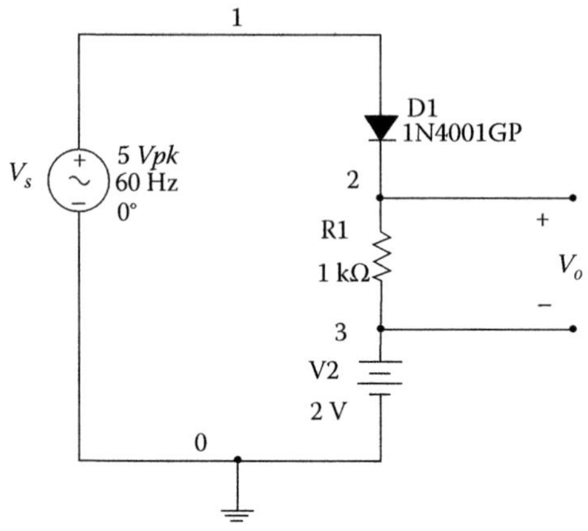

FIGURE 6.10
Battery-charging circuit.

Solution:

Build the Circuit:

Use a breadboard to build the circuit shown in Figure 6.10. Connect the AWG1 (yellow wire) to node 1 of the circuit. Also, connect the scope channel 1 positive (orange wire) to node 1. In addition, connect the ground (black wire) to node 0 of the circuit. Moreover, connect the scope channel 1 negative (orange-white wire) to node 0. Furthermore, connect the scope channel 2 positive (blue wire) to node 2, and the scope channel 2 negative (blue-white wire) to node 3.

Generation of Sine Wave:

Click on the WaveGen in the main WaveForms screen. Go to Basic → Sine → Frequency (set to 60 Hz) → Amplitude (set to 5 V) → Offset (set to 0 V) → Symmetry (set to 50%) → Click Run AWG1.

Generation of the DC Input Voltage

Select Channels → Channel 2 (AWG2) → Go to Basic → dc → 2 V → Click Run AWG2.

Activation of Scope:

Click on the Scope in the main WaveForms screen. Go to Run → Autoset → Measure → Add → Channel 1 → Horizontal → Frequency → Add Selected Measurement → Add → Channel 1 → Vertical → Peak to Peak → Add Selected Instrument → Add → Channel 2 → Vertical → Peak to Peak → Control → Add Mathematic Channel → Custom → Type C2/1000 (this current is calculated as the voltage across the resistor 1000 Ω divided by resistance 1000 Ω) → control → Change Math Mode Units to A → Change the M1 range to 2 mA/V.

Figure 6.11 shows the Digilent Analog Discovery WaveGen setup, and Figure 6.12 shows the oscilloscope display of the input and output waveforms.

FIGURE 6.11
Analog Discovery WaveGen setup for battery-charging circuit.

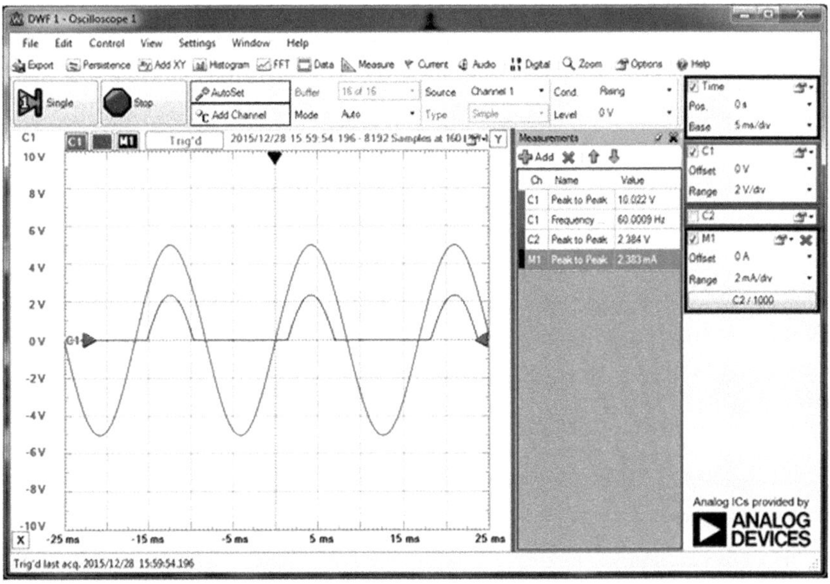

FIGURE 6.12
The current flowing through the diode (scope channel M1).

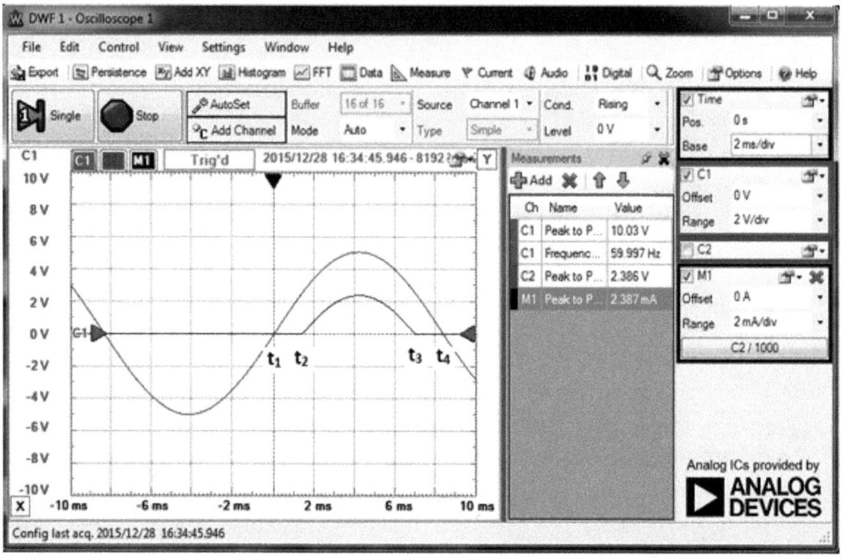

FIGURE 6.13
Conduction angle of a battery-charging circuit.

From Figure 6.13, the following times were measured:

t_1=0.1 ms,
t_2=1.5 ms,
t_3=7.0 ms, and
t_4=8.4 ms

The conduction angle is that part of the cycle during which the diode is conducting.

$$\text{Conduction angle} = \frac{t_3 - t_2}{2(t_4 - t_1)} * 360 \tag{6.9}$$

Using Equation 6.9 and the values of t_1, t_2, t_3, and t_4 obtained from Figure 6.13, the conduction angle is

$$\text{Conduction angle} = 119.28°$$

From Figure 6.13, the peak current flowing through the diode is 2.387 mA.

6.3 Peak Detector

A peak detector is a circuit that can be used to detect the peak value of an input signal. Peak detectors can also be used as demodulators to detect the

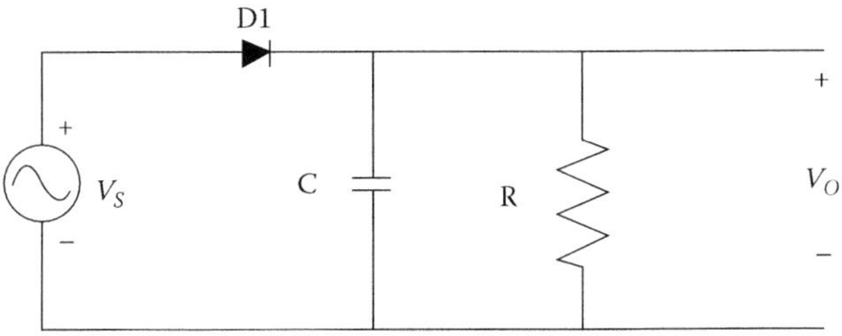

FIGURE 6.14
Peak detector circuit.

audio signal in amplitude modulated (AM) waves. A simple peak detector circuit is shown in Figure 6.14.

The peak detector circuit is simply a half-wave rectifier circuit with a capacitor connected across the load resistor. The operation of the circuit will be described with the assumption that the source voltage is a sinusoidal voltage, $V_m \sin(2\pi f_o t)$, with amplitude greater than 0.6 volts. During the first quarter-cycle, the input voltage increases and the capacitor is charged to the input voltage. At time $t = 1/4f_o$, where f_o is the frequency of the sinusoidal input, the input voltage reaches its maximum value of V_m and the capacitor will be charged to that maximum of V_m.

When time $t > 1/4f_o$, the input starts to decrease; the diode D1 will discharge through the resistance R. If we define $t_1 = 1/4f_o$, the time when the capacitor is charged to the maximum value of the input voltage, the discharge of the capacitor is given as

$$v_0(t) = V_m e^{-(t-t_1)/RC} \tag{6.10}$$

If $RC >> T$, where T is the period of the input signal, it can be shown that the ripple voltage can be given as

$$V_{r(peak-to-peak)} = \frac{V_m}{f_0 CR} \tag{6.11}$$

where f_o is the frequency of the input ac source voltage.

Example 6.4 demonstrates the effects of the time constant on the characteristics of a peak detector circuit.

Example 6.4 Ripple Voltage of Half-Wave Rectifier

Figure 6.15 shows a peak detector. (1) Display the input and output voltages. (2) Measure the ripple voltage for load resistors of 10, 50, and

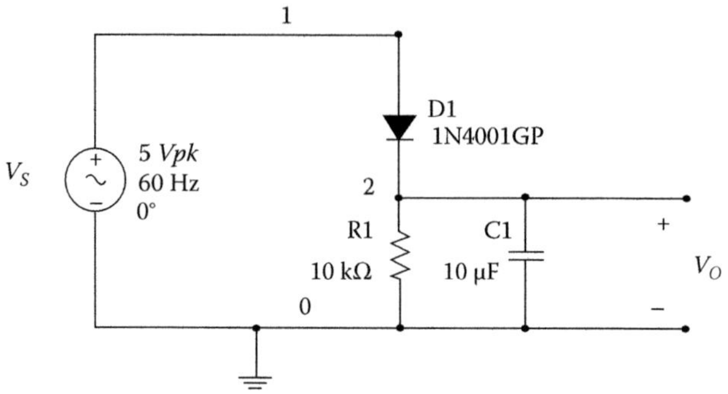

FIGURE 6.15
Peak detector circuit for measuring ripple voltage.

100 kΩ. (3) Use Equation 6.11 to calculate the ripple voltages for load resistances of 10, 50, and 100 kΩ and compare the calculated values with the measured ones.

Solution:

Build the Circuit:

Use a breadboard to build the circuit shown in Figure 6.15. Connect the AWG1 (yellow wire) to node 1 of the circuit. Also, connect the scope channel 1 positive (orange wire) to node 1. In addition, connect the scope channel 1 negative (orange-white wire) to node 0. Moreover, connect the ground (black wire) to node 0 of the circuit. Furthermore, connect the scope channel 2 positive (blue wire) to node 2, and the scope channel 2 negative (blue-white wire) to node 0.

Generation of Sine Wave:

Click on the WaveGen in the main WaveForms screen. Go to Basic → Sine → Frequency (set to 60 Hz) → Amplitude (set to 5 V) → Offset (set to 0 V) → Symmetry (set to 50%) → Click Run AWG1. Figure 6.16 shows the setup for Analog Discovery WaveGen AWG1.

Activation of Scope:

Click on the Scope in the main WaveForms screen. Go to Run → Autoset → Measure → Add → Channel 1 → Horizontal → Frequency → Add Selected Measurement → Add → Channel 1 → Vertical → Peak to Peak → Add Selected Instrument → Add → Channel 2 → Vertical → Peak to Peak → Add Selected Measurement → Close Add Instrument. Figure 6.17 shows the oscilloscope display of the input and output waveforms.

For each of the load resistors of 10, 50, and 100 kΩ, the ripple voltage was measured. In addition, Equation 6.11 was used to calculate the ripple voltage. Table 6.1 shows the calculated and measured ripple voltages. It can be seen from Table 6.1 that there are agreements between the calculated and measured ripple voltages, especially for the 50 kΩ resistor.

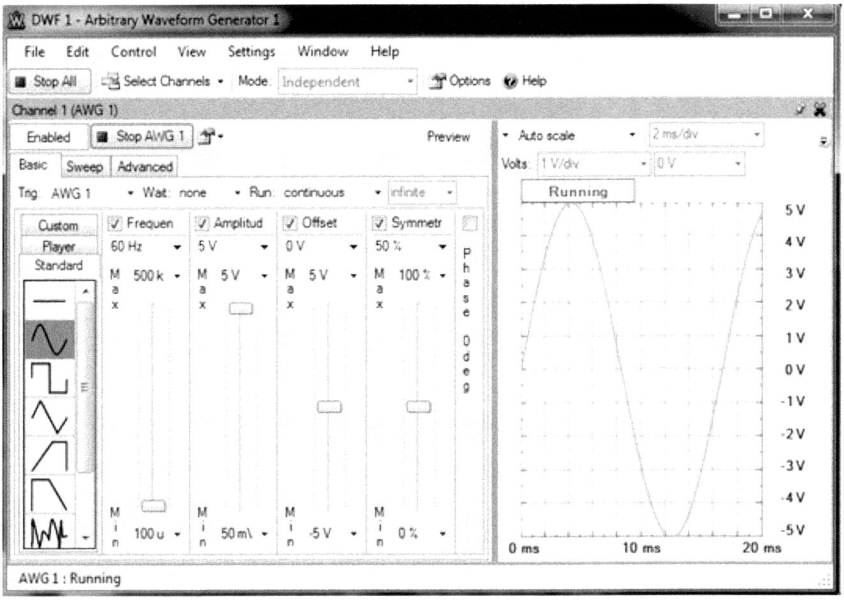

FIGURE 6.16
Analog Discovery WaveGen setup for peak detector circuit.

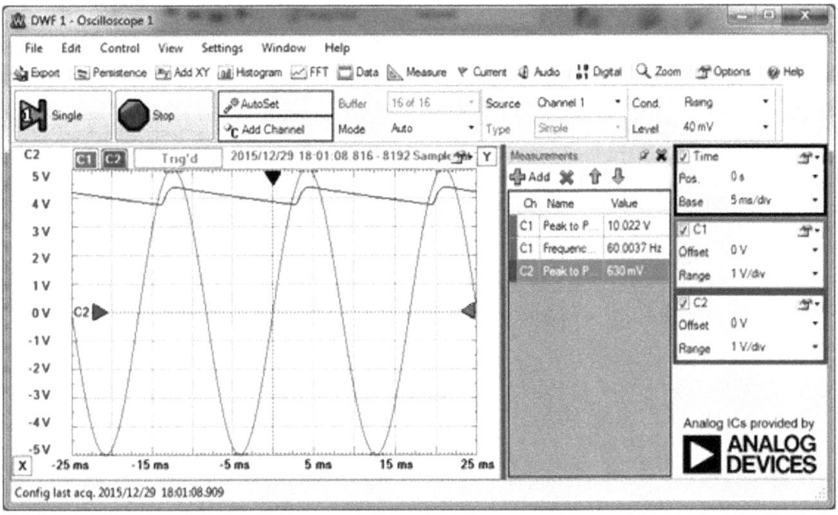

FIGURE 6.17
Display of input (scope channel 1) and output (scope channel 2) waveforms of a peak detector with 10 kΩ resistor.

TABLE 6.1
Measured and Calculated Ripple Voltages for Peak Detector Circuit

Resistance (Ω)	Capacitance (F)	Frequency (Hz)	Peak Voltage (V) (5 V-Diode Drop)	Calculated Ripple Voltage (V)	Measured Ripple Voltage (V)
1.00E+04	1.00E–05	60	4.3	7.17E–01	6.34E–01
5.00E+04	1.00E–05	60	4.3	1.43E–01	1.54E–01
1.00E+05	1.00E–05	60	4.3	7.17E–02	8.40E–02

6.4 Full-Wave Rectification

A full-wave rectifier that does not require a center-tapped transformer is the bridge rectifier of Figure 6.18. When $v_S(t)$ is positive, diodes D1 and D3 conduct, but diodes D2 and D4 do not conduct. The current enters the load resistor R through node A. In addition, when $v_S(t)$is negative, the diodes D2 and D4 conduct, but diodes D1 and D3 do not conduct. The current entering the load resistor R enters through node A. The output voltage is

$$v(t) = |v_S(t)| - 2V_D \tag{6.12}$$

Example 6.5 explores full-wave rectification with real diodes.

Example 6.5 Full-Wave Bridge Rectifier

For the bridge rectifier shown in Figure 6.19, use the Analog Discovery board to display the input and output waveforms. What is the peak-to-peak voltage at the output of the bridge rectifier?

Solution:
Build the Circuit:
Use a breadboard to build the circuit shown in Figure 6.19. Connect the AWG1 (yellow wire) to node 1 of the circuit. Also, connect the scope channel 1 positive (orange wire) to node 1. In addition, connect

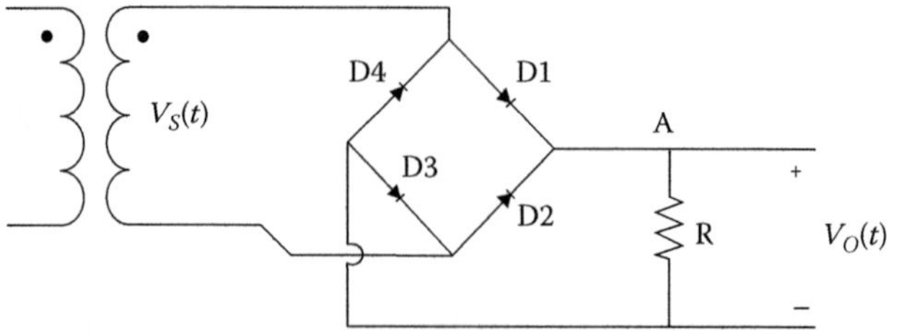

FIGURE 6.18
Bridge rectifier.

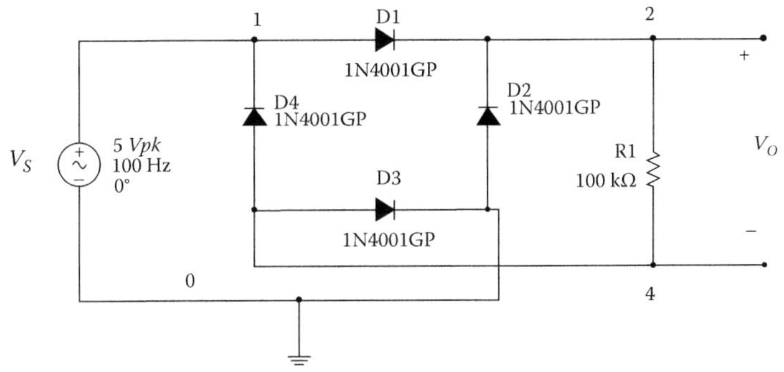

FIGURE 6.19
Full-wave bridge rectifier circuit built using diodes 1N4001.

the scope channel 1 negative (orange-white wire) to node 0. Moreover, connect the ground (black wire) to node 0 of the circuit. Furthermore, connect the scope channel 2 positive (blue wire) to node 2, and the scope channel 2 negative (blue-white wire) to node 0.

Generation of Sine Wave:
Click on the WaveGen in the main WaveForms screen. Go to Basic → Sine → Frequency (set to 100 Hz) → Amplitude (set to 5 V) → Offset (set to 0 V) → Symmetry (set to 50%) → Click Run AWG1. Figure 6.20 shows the setup of the Analog Discovery WaveGen AWG1.

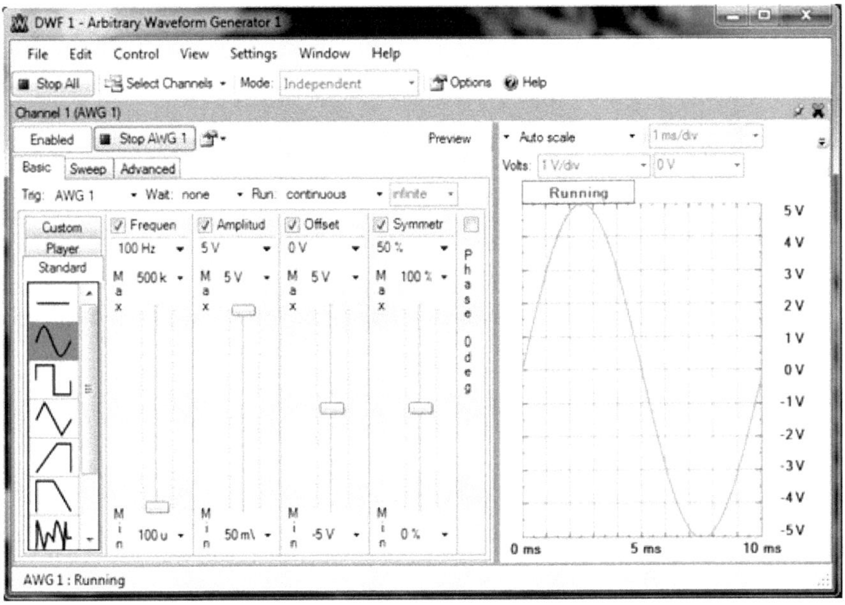

FIGURE 6.20
Analog Discovery WaveGen setup for full-wave rectifier circuit.

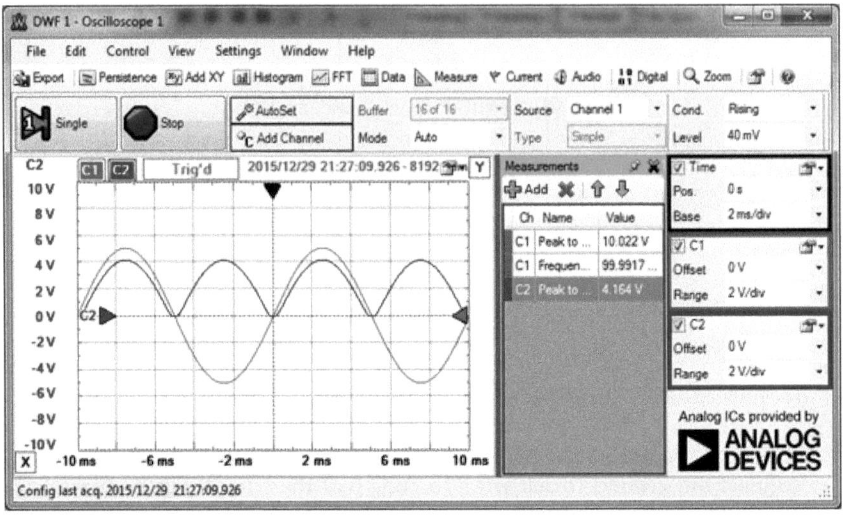

FIGURE 6.21
Input (scope channel 1) and output (scope channel 2) waveforms of a full-wave rectifier.

Activation of Scope:
Click on the Scope in the main WaveForms screen. Go to Run → Autoset → Measure → Add → Channel 1 → Horizontal → Frequency → Add Selected Measurement → Add → Channel 1 → Vertical → Peak to Peak → Add Selected Instrument → Add → Channel 2 → Vertical → Peak to Peak → Add Selected Instrument → Close Add Instrument. Figure 6.21 shows the oscilloscope display of the full-wave rectification.

6.5 Full-Wave Rectifier with Smoothing Filter

A bridge rectifier with RC smoothing filter is shown in Figure 6.22. Connecting a capacitor across the load of a bridge rectifier can smooth the output voltage

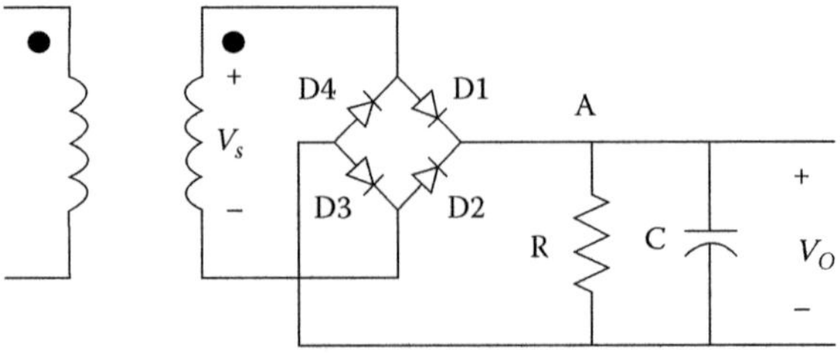

FIGURE 6.22
Bridge rectifier with RC smoothing filter.

of a full-wave rectifier circuit by filtering out the high-frequency components of the output signal.

For full-wave rectification, the frequency of the ripple voltage is twice that of the input voltage. The capacitor in Figure 6.22 has only half the time to discharge. Therefore, for a given time constant, *CR*, the ripple voltage will be reduced, and it is given by

$$V_{r(peak-to-peak)} = \frac{V_m}{2f_oCR} \tag{6.13}$$

where

V_m is peak value of the input sinusoidal waveform
f_o is the frequency of the input sinusoidal waveform

and the average dc voltage at the output is approximately

$$V_{dc} = V_m - \frac{V_r}{2} = V_m - \frac{V_m}{4f_oCR} \tag{6.14}$$

Example 6.6 demonstrates the effects of the RC smoothing filter on the output of Figure 6.22.

Example 6.6 Full-Wave Rectification with a Smoothing Filter

Figure 6.23 shows a bridge rectifier circuit with a smoothing filter. (1) Display the input and output voltages. (2) Measure the ripple voltage for load capacitors of 0.22, 0.44, and 0.66 μF. (3) Calculate the ripple voltages for load capacitances of 0.22, 0.44, and 0.66 μF and compare the calculated values with the measured ones.

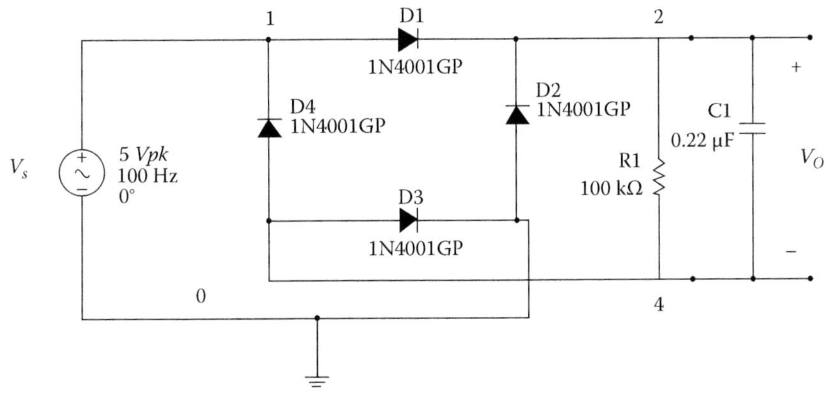

FIGURE 6.23
Bridge rectifier with RC smoothing filter.

Solution:

Build the Circuit:

Use a breadboard to build the circuit shown in Figure 6.23. Connect the AWG1 (yellow wire) to node 1 of the circuit. Also, connect the scope channel 1 positive (orange wire) to node 1. In addition, connect the scope channel 1 negative (orange-white wire) to node 0. Moreover, connect the ground (black wire) to node 0 of the circuit. Furthermore, connect the scope channel 2 positive (blue wire) to node 2, and the scope channel 2 negative (blue-white wire) to node 0.

Generation of Sine Wave:

Click on the WaveGen in the main WaveForms screen. Go to Basic → Sine → Frequency (set to 100 Hz) → Amplitude (set to 5 V) → Offset (set to 0 V) → Symmetry (set to 50%) → Click Run AWG1. Figure 6.24 shows the setup of the Analog Discovery WaveGen AWG1.

Activation of Scope:

Click on the Scope in the main WaveForms screen. Go to Run → Autoset → Measure → Add → Channel 1 → Horizontal → Frequency → Add Selected Measurement → Add → Channel 1 → Vertical → Peak to Peak → Add Selected Instrument → Add → Channel 2 → Vertical → Peak to Peak → Add Selected Instrument → Close Add Instrument. Figure 6.25 shows the oscilloscope display of the input and output waveforms.

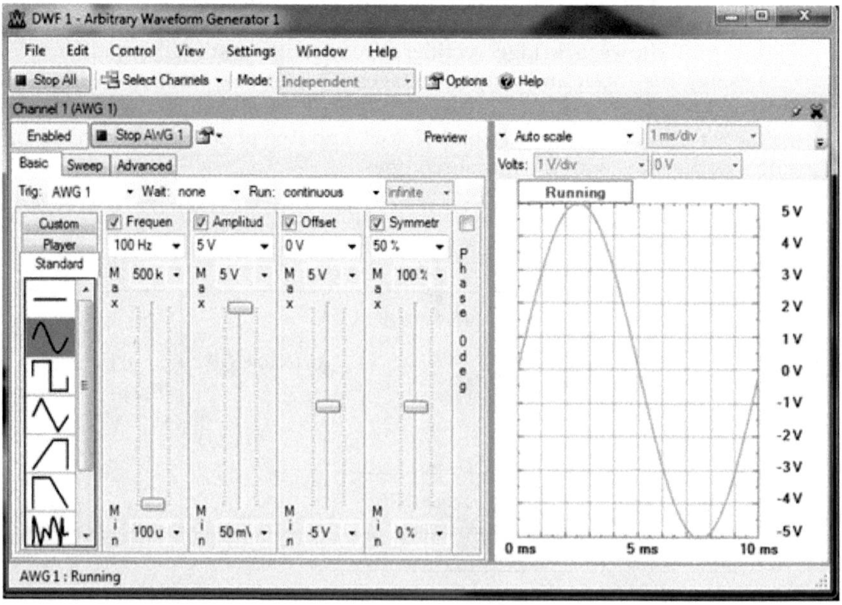

FIGURE 6.24
Analog Discovery WaveGen setup for bridge rectifier with RC filter.

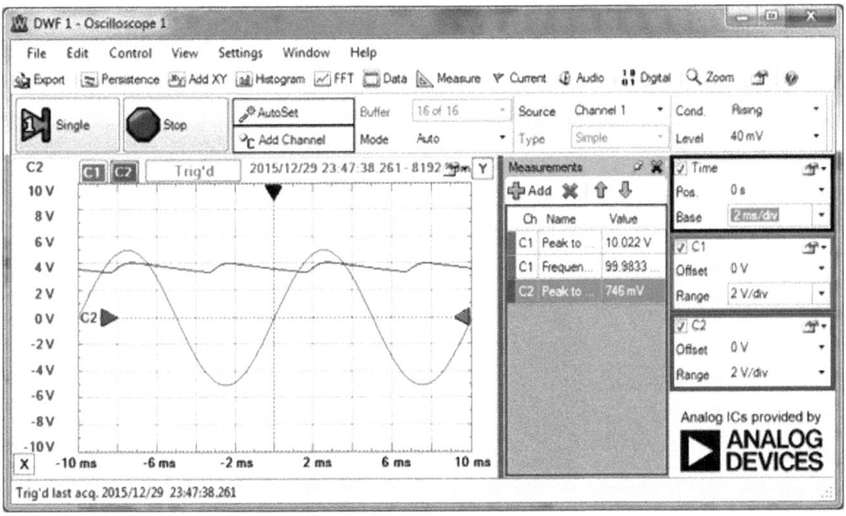

FIGURE 6.25
Input (scope channel 1) and output (scope channel 2) waveforms of bridge rectifier with RC smoothing filter.

TABLE 6.2
Measured and Calculated Ripple Voltages for Bridge Rectifier with RC Smoothing Filter

Resistance (Ω)	Capacitance (F)	Frequency (Hz)	Peak Voltage (V) (5 V-2 Diode Drops)	Calculated Ripple Voltage (V)	Measured Ripple Voltage (V)
1.00E+05	2.20E–07	100	3.6	8.18E–01	6.90E–01
1.00E+05	4.40E–07	100	3.6	4.09E–01	4.02E–01
1.00E+05	6.60E–07	100	3.6	2.73E–01	2.86E–01

For each of the load capacitors of 0.22, 0.44, and 0.66 μF, the ripple voltage was measured. In addition, Equation 6.13 was used to calculate the ripple voltage. Table 6.2 shows the calculated and measured ripple voltages. It can be seen from Table 6.2 that there are agreements between the calculated and measured ripple voltages, especially for the 0.44 μF capacitor.

PROBLEMS

Problem 6.1 Determine the I–V characteristics of a green LED in the circuit shown in Figure P6.1.

Problem 6.2 Obtain the I–V characteristics of a red LED in the circuit shown in Figure P6.2.

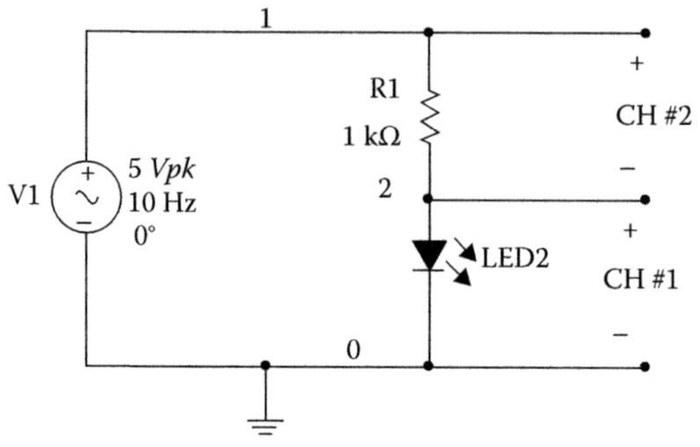

FIGURE P6.1
Resistive circuit with green LED.

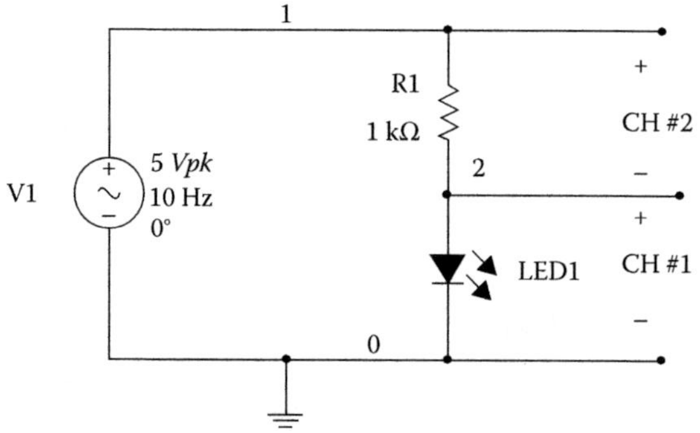

FIGURE P6.2
Resistive circuit with red LED.

Problem P6.3 A battery-charging circuit is shown in Figure P6.3. The battery voltage is 3 V. The source voltage is $v_S(t) = 5\sin(120\pi t)$ V, and the load resistance is 1000 Ω. Use the Analog Discovery board to (1) display the input voltage, (2) plot the current flowing through the diode, (3) measure the conduction angle of the diode, and (4) determine the peak current flowing through the diode.

Problem P6.4 A battery-charging circuit is shown in Figure P6.4. The battery voltage is 1 V. The source voltage is $v_S(t) = 5\sin(120\pi t)$ V, and the load resistance is 500 Ω. Use the Analog Discovery board to (1) display the input voltage, (2) plot the current flowing through the diode, (3) measure the conduction angle of the diode, and (4) determine the peak current flowing through the diode.

Problem P6.5 A half-wave rectifier, built using red LED, is shown in Figure P6.5. (1) Use the Analog Discovery board to display the input and output voltages. (2) Calculate the maximum current flowing through the LED.

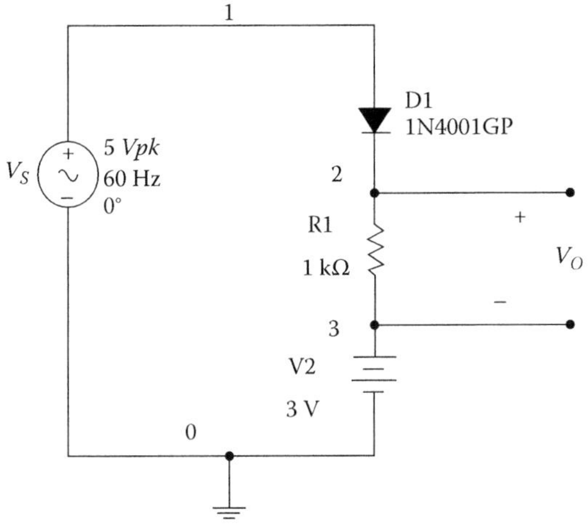

FIGURE P6.3
Battery-charging circuit with battery voltage of 3 V.

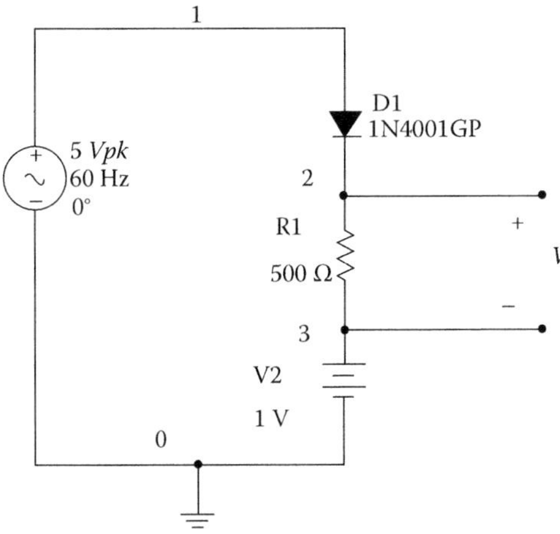

FIGURE P6.4
Battery-charging circuit with battery voltage of 1 V.

(3) If the frequency of the source is increased to 100 Hz, why is the switching on and off of the LED not visible to the eye?

Problem P6.6 Figure P6.6 shows a peak detector. (1) Display the input and output voltages. (2) Measure the ripple voltage for load resistors of source frequencies of 60, 600, and 6000 Hz. (3) Calculate the ripple voltages for source frequencies of 60, 600, and 6000 Hz and compare the calculated values with the measured ones.

Problem P6.7 A full-wave rectifier, built using LEDs and diodes, is shown in Figure P6.7. LED1 is a green LED, and LED2 is a red LED. (1) Use the Analog

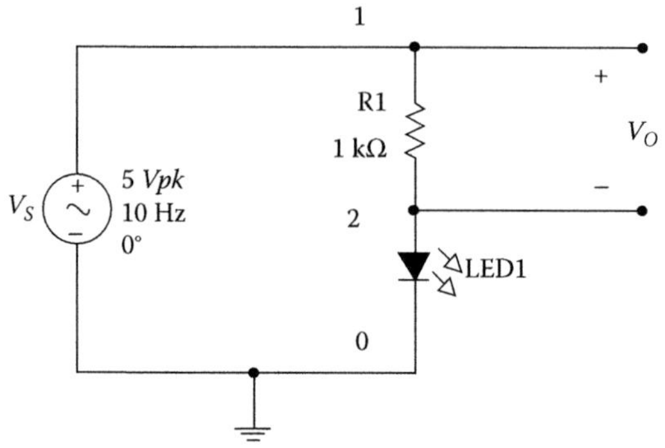

FIGURE P6.5
Half-wave rectifier with LED.

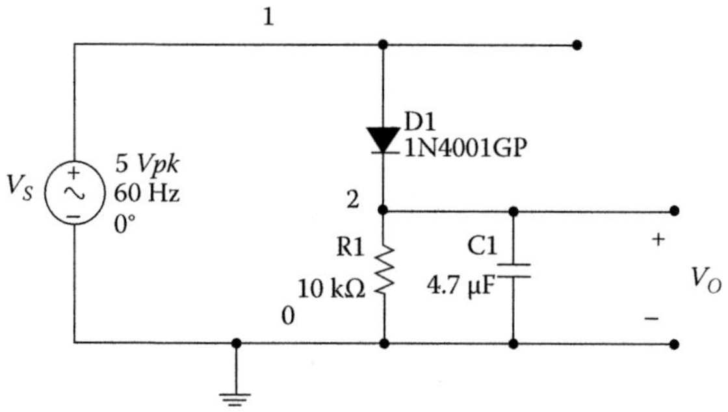

FIGURE P6.6
Peak detector for measuring the ripple voltage at various frequencies.

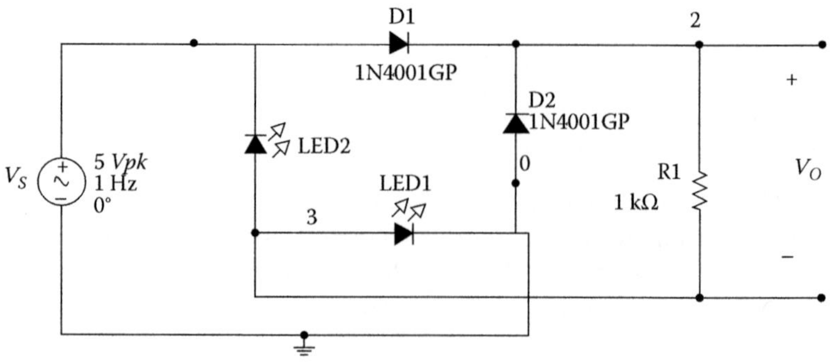

FIGURE P6.7
Bridge rectifier with LEDs and diodes.

Discovery board to display the input and output voltages. (2) What is the peak-to-peak voltage across the resistor R1? (3) Why do the two LEDs alternately turn on and off? (4) If the frequency of the source is increased to 100 Hz, why would the switching on and off of the LEDs not be visible to the eye?

Problem P6.8 A full-wave rectifier, built using LEDs, diodes, and an RC filter, is shown in Figure P6.8. LED1 is a green LED, and LED2 is a red LED. (1) Use the Analog Discovery board to display the input and output voltages. (2) What is the peak-to-peak voltage across the resistor R1?

Problem 6.9 Figure P6.9 shows a bridge rectifier circuit with a smoothing filter. (1) Display the input and output voltages. (2) Measure the ripple voltage for source frequencies of 60, 120, and 180 Hz. (3) Calculate the ripple voltages for source frequencies of 60, 120, and 180 Hz and compare the calculated values with the measured ones.

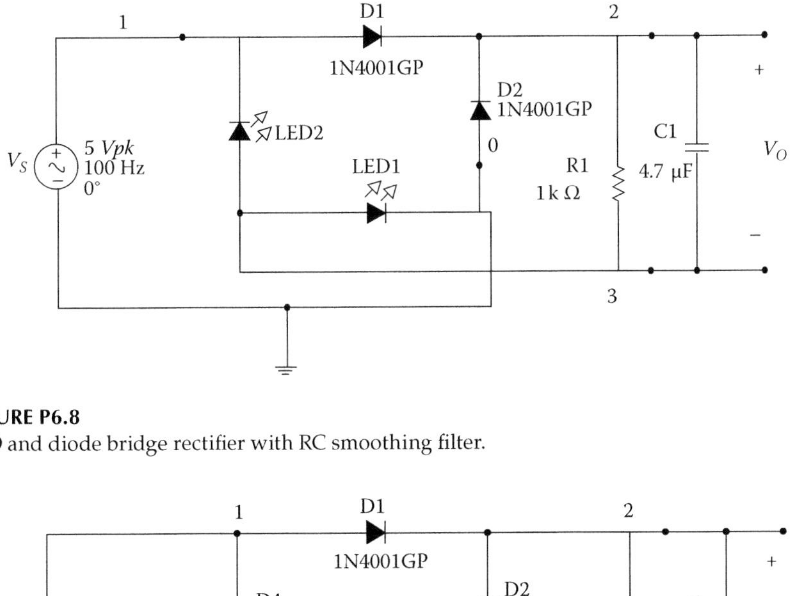

FIGURE P6.8
LED and diode bridge rectifier with RC smoothing filter.

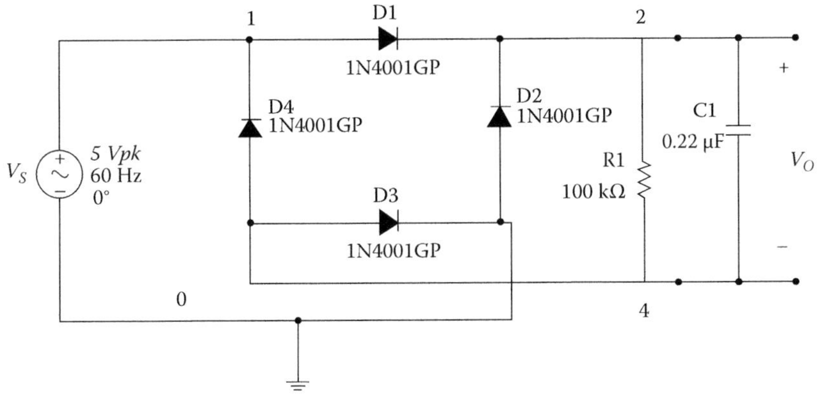

FIGURE P6.9
Bridge rectifier with RC smoothing filter.

7

Transistors

INTRODUCTION In this chapter, Analog Discovery board will be used to solve problems involving metal-oxide semiconductor field effect and bipolar junction transistors. The general topics to be discussed in this chapter are dc model of BJT and MOSFET, biasing of discrete circuits, and frequency response of amplifiers.

7.1 Bipolar Junction Transistors

Bipolar junction transistor (BJT) consists of two pn junctions connected back to back. There are two types of BJT: NPN and PNP transistors. The electronic symbols of the two types of transistors are shown in Figure 7.1.

The dc behavior of the BJT can be described by the Ebers-Moll model. The voltages of the base-emitter and base-collector junctions define the region of operation of the BJT. The regions of normal operation are forward-active, reverse-active, saturation, and cut-off. Table 7.1 shows the regions of operation based on the polarities of the base-emitter and base-collector junctions. The regions of operation are described below.

1. Forward-Active Region

   The forward-active region corresponds to forward biasing the emitter-base junction and reverse biasing the base-collector junction. It is the normal operation region of bipolar junction transistors when employed for amplifications. In the forward-active region, the first-order representations of collector current I_C and base current I_B are given as

$$I_C = I_S \exp\left(\frac{V_{BE}}{V_T}\right)\left(1 + \frac{V_{CE}}{V_{AF}}\right) \tag{7.1}$$

   and

$$I_B = \frac{I_S}{\beta_F} \exp\left(\frac{V_{BE}}{V_T}\right) \tag{7.2}$$

   where

   β_F is large signal forward current gain of common-emitter configuration,

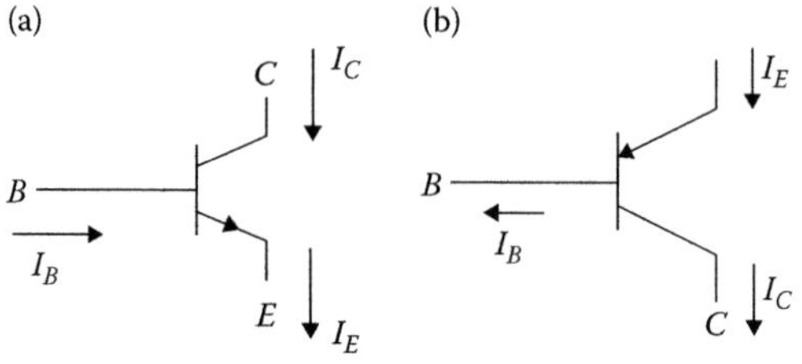

FIGURE 7.1
(a) NPN transistor and (b) PNP transistor.

TABLE 7.1
Regions of Operation of BJT

Base-Emitter Junction	Base-Collector Junction	Region of Operation
Forward-biased	Reverse-biased	Forward-active
Forward-biased	Forward-biased	Saturation
Reverse-biased	Reverse-biased	Cut-off
Reverse-biased	Forward-biased	Reverse-active

V_{AF} is forward early voltage,
I_S is the BJT transport saturation current.
V_T is the thermal voltage and is given as

$$V_T = \frac{kT}{q} \tag{7.3}$$

where
k is the Boltzmann's Constant ($k = 1.381 \times 10^{-23}$ V C/°K),
T is the absolute temperature in degrees Kelvin,
q is the charge of an electron ($q = 1.602 \times 10^{-19}$ C).

If $V_{AF} >> V_{CE}$, then from Equations 7.1 and 7.2, we have

$$I_C = \beta_F I_B \tag{7.4}$$

2. Reverse-Active Region

The reverse-active region corresponds to reverse biasing the emitter-base junction and forward biasing the base-collector junction. The Ebers-Moll model in the reverse-active region ($V_{BC} > 0.5$ V and $V_{BE} < 0.3$ V) simplifies to

$$I_E = I_S\left[\frac{V_{BC}}{V_T}\right] \tag{7.5}$$

$$I_B = \frac{I_S}{\beta_R} \exp\left[\frac{V_{BC}}{V_T}\right] \tag{7.6}$$

Thus

$$I_E = \beta_R I_B$$

The reverse-active region is seldom used.

3. Saturation Region

   The saturation region corresponds to forward biasing both base-emitter and base-collector junctions. A switching transistor will be in the saturation region when the device is in the conducting or "ON" state. Equation 7.4 is not valid when the transistor is operating in the saturation region. In the saturation region and other regions of operation, one can use Kirchhoff's Current Law to obtain the expression

$$I_E = I_C + I_B \tag{7.7}$$

4. Cut-off Region

   The cut-off region corresponds to reverse biasing the base-emitter and base-collector junctions. The collector and base currents are very small compared to the currents that flow into the collector or base when transistors are in the active-forward and saturation regions. In most applications, it is adequate to assume that $I_C = I_B = I_E = 0$ when a BJT is in the cut-off region. A switching transistor will be in the cut-off region when the device is not conducting or in the "OFF" state.

From Equation 7.2, the input characteristic of a forward biased base-emitter junction is similar to diode characteristics. In Example 7.1 and Example 7.2, we explore the input and output characteristics of a BJT transistor.

Example 7.1 BJT Output Characteristics

Use the Analog Discovery board to determine the current vs. voltage (I_C vs. V_{CE}) curve of the NPN transistor 2N3904.

Solution:

Build the Circuit:

Use a breadboard to build the circuit shown in Figure 7.2. Connect the AWG1 (yellow wire) to node 3 of the circuit and the AWG2 (yellow-white wire) to node 1 of the circuit. Also, connect the scope channel 1 positive (orange wire) to node 1. In addition, connect the scope channel 1 negative (orange-white wire) to node 0. Moreover, connect the ground (black wire) to node 0 of the circuit. Furthermore, connect the scope channel 2

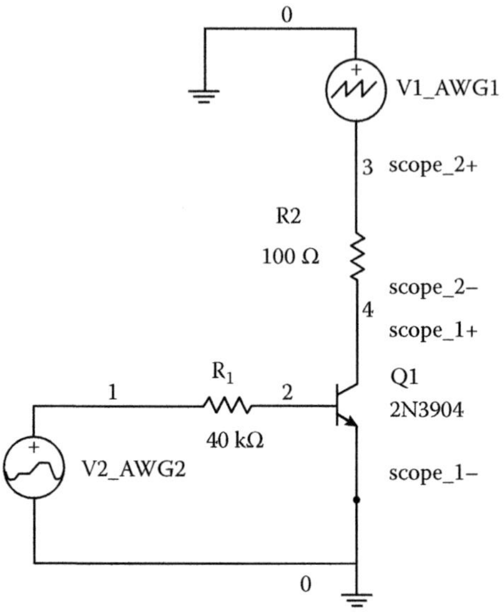

FIGURE 7.2
Circuit for determining current vs. voltage characteristics of 2N3904.

positive (blue wire) to node 3, and the scope channel 2 negative (blue-white wire) to node 4.

The pin configuration of transistor 2N3904 is shown in Figure 7.3. For the transistor 2N3904, connect the base of the transistor to node 2, collector to node 4, and the emitter to node 0.

Generation of the Triangular Waveform of AWG1:
AWG1 generates collector voltage, V_{CC}. The triangular waveform is made to range from 0 to 5 V. Each ramp generates an I–V curve for a specific value of the base current, I_B. The triangular waveform is generated

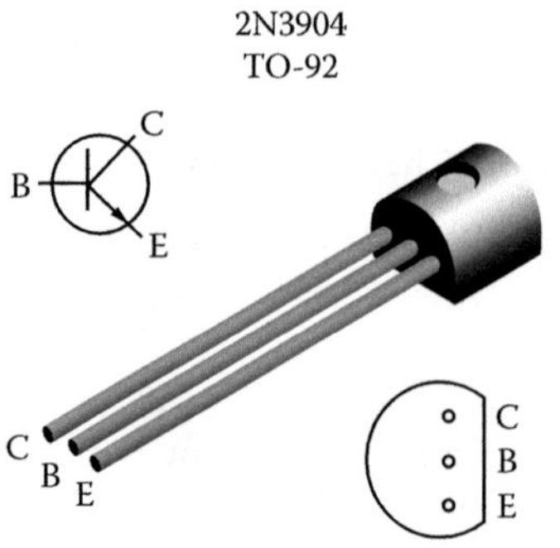

FIGURE 7.3
Pin configuration of 2N3904 transistor. (From Fairchild Datasheets, www.fairchildsemi.com/.)

through the following commands: Click on the WaveGen in the main WaveForms screen. Go to Basic → Triangular → Frequency (set to 50 Hz) → Amplitude (set to 2.5 V) → Offset (set to 2.5 V) → Symmetry (set to 50%) → Click Run AWG1.

Generation of the Step Function of AWG2:
AWG2 generates uniformly distributed steps (stairs waveform). To generate the stairs waveform, open up an Excel file and insert the numbers 1, 2, 3, 4, …, 10 in ten rows. The Excel file should be saved with either .csv or .txt extensions.

Open the Arbitrary Waveform Generator, AWG2, Go to Custom → File → Select source file → Import data in source file with the file extension .csv or .txt → Frequency (set to 5 Hz, frequency is the buffer iteration frequency, and 5 Hz means the whole stair sequence takes 200 ms, or 20 ms, per step) → Amplitude (set to 2.5 V) Offset (set to 2.5 V) → Symmetry (set to 50%) → Click Run AWG2. Figure 7.4 shows the waveforms generated by AWG1 and AWG2.

Activation of Scope:
Scope channel 1 measures the collector and emitter voltage (V_{CE}). Scope channel 2 measures the voltage across the resistor R2 (100 Ω). To obtain the collector current, we calculate C2/R2. The Mathematics Channel is used to obtain the collector current. Follow these steps to obtain the

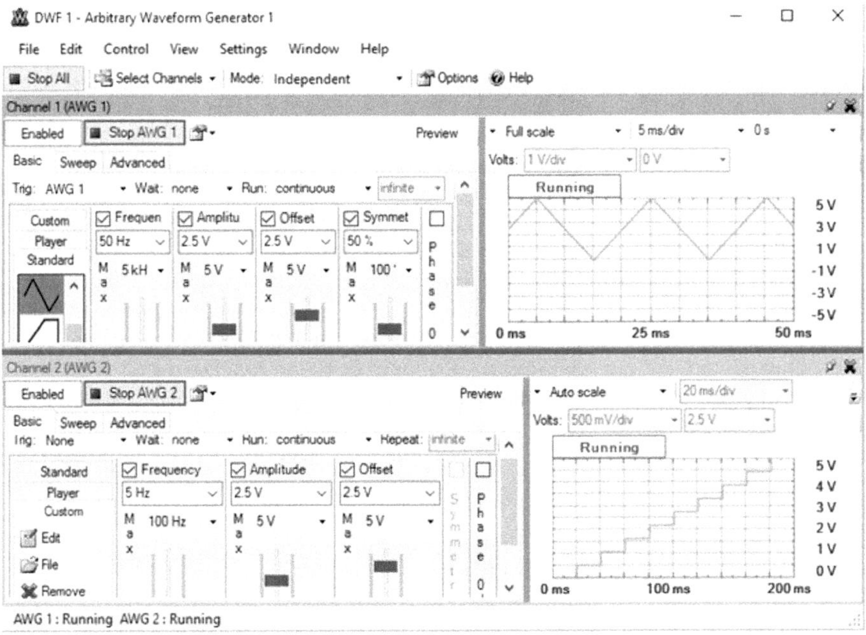

FIGURE 7.4
Arbitrary waveform generator 1 and 2 (AWG1 and AWG2) settings.

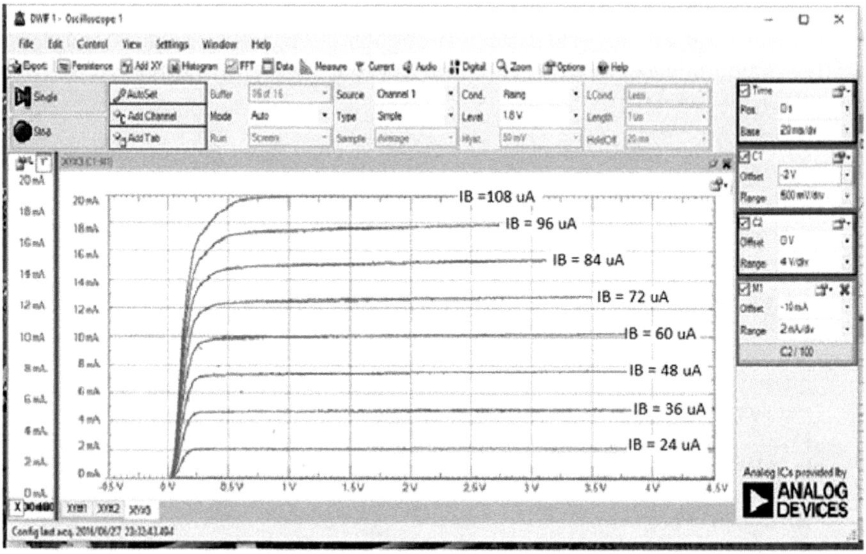

FIGURE 7.5
Output current vs. voltage characteristics of 2N3904 transistor.

I–V curves: Add Mathematic Channel → Custom → Type C2/100 (this current is calculated as the voltage across the 100 Ω resistor divided by resistance of 100 Ω) → control → Add XY (this initiates the XY plot) → set X to C1 (Channel 1) and set Y to M1 (Mathematic Channel 1) → Change Math Mode Units to A → Change the M1 range to 1 mA/V. Figure 7.5 shows the output characteristics (I_C vs. V_{CE}) of the transistor 2N3904.

With regard to the output I–V curve of the BJT, the maximum voltage at the base is 5 V. The base resistor has a value of 40 K, and assuming a 0.7 V drop across the base-emitter junction, the maximum base current is (5 – 0.7)/40 K = 0.1075 mA. Since 10 steps were used in generation of the base current, the step current at the base is a (0.1075/9) mA, which is approximately equal to 0.012 mA. It should be noted that the diode drop is assumed to be 0.7 V, so the base currents shown on the I–V curve are approximate values.

Example 7.2 BJT Input Characteristics

Use the Analog Discovery board to determine the input current vs. voltage (I_B vs. V_{BE}) curve of the NPN transistor 2N3904.

Solution:

Use a breadboard to build the circuit shown in Figure 7.6. Connect the AWG1 (yellow wire) to node 3 of the circuit and the AWG2 (yellow-white wire) to node 1 of the circuit. Also, connect the scope channel 1

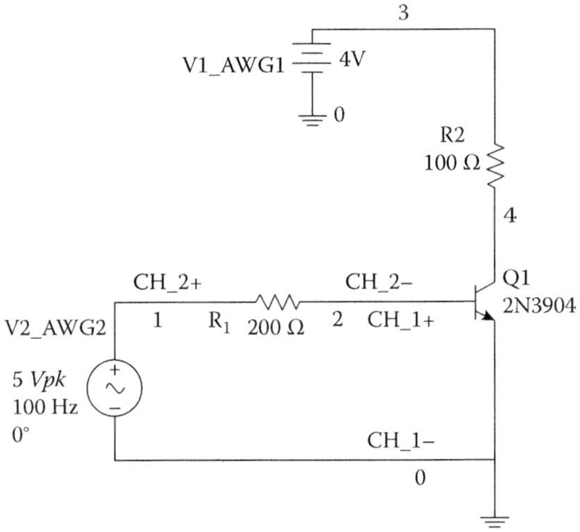

FIGURE 7.6
Circuit for determining input current vs. voltage characteristics of 2N3904.

positive (orange wire) to node 2. In addition, connect the scope channel 1 negative (orange-white wire) to node 0. Furthermore, connect the ground (black wire) to node 0 of the circuit. Moreover, connect the scope channel 2 positive (blue wire) to node 1, and the scope channel 2 negative (blue-white wire) to node 2.

For the transistor 2N3904, connect the base of the transistor to node 2, the collector to node 4, and the emitter to node 0 of Figure 7.6.

Generation of DC Voltage:
Click on the WaveGen 1 (AWG1) in the main WaveForms screen. Go to Basic → dc → Offset (set to 4 V) → Click Run AWG1.

Generation of Sine Wave:
Click on the WaveGen 2 (AWG2) in the main WaveForms screen. Go to Basic → Sine → Frequency (set to 100 Hz) → Amplitude (set to 5 V) → Offset (set to 0 V) → Symmetry (set to 50%) → Click Run AWG2. Figure 7.7 shows the setup of the Analog Discovery WaveGen AWG1 and AWG2.

Activation of Scope:
Click on the Scope in the main WaveForms screen. Go to Run → Autoset → Measure → Add → Channel 1 → Horizontal → Frequency → Add Selected Measurement → Add → Channel 1 → Vertical → Peak to Peak → Add Selected Instrument → Add → Channel 2 → Vertical → Peak to Peak → Control → Add Mathematic Channel → Custom → Type C2/200 (This current is calculated as the voltage across the resistor 200 Ω divided by resistance 200 Ω) → control → Add XY (this initiates the XY plot) → set X (Channel 1) and set Y (Math1) → Change Math Mode Units to A → Change the M1 range to 1 mA/V. Figure 7.8 shows the input I–V characteristics of the transistor 2N3904.

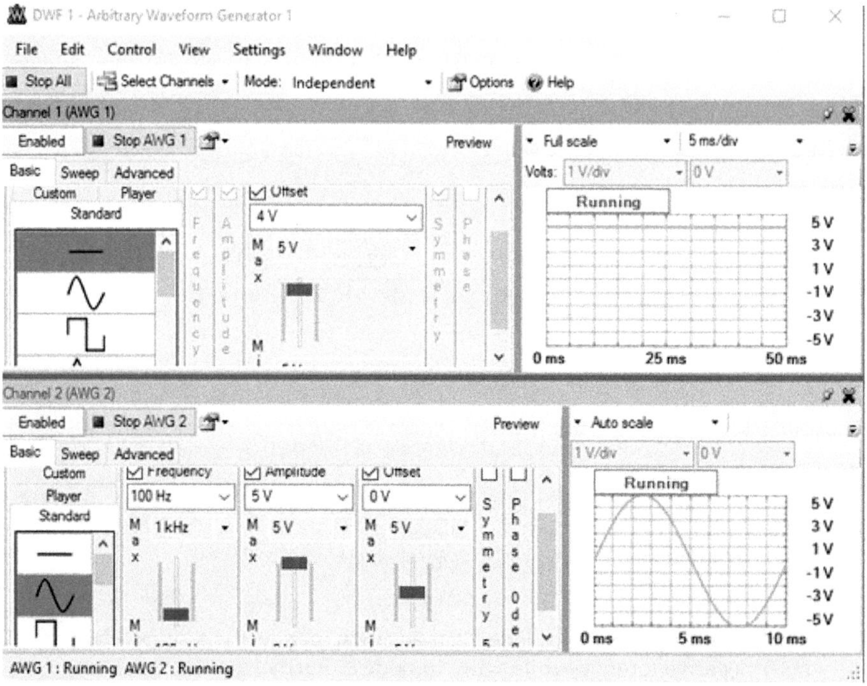

FIGURE 7.7
The analog discovery WaveGen setup.

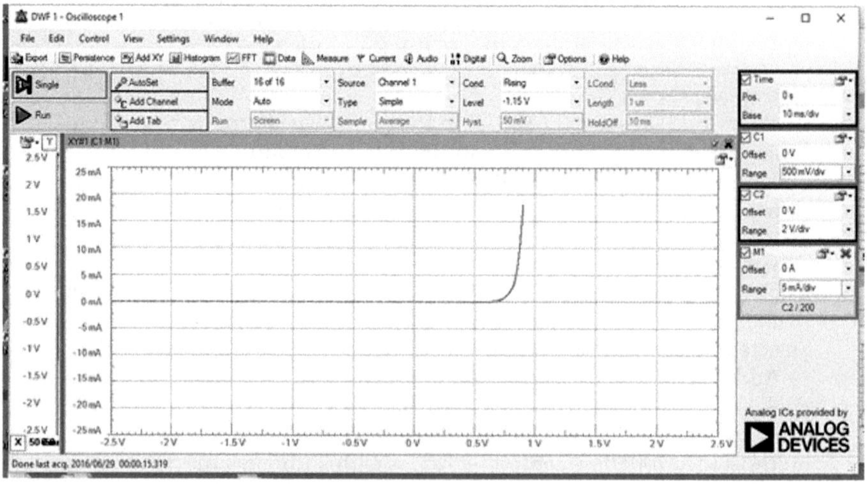

FIGURE 7.8
The input I–V (I_B vs. V_{BE}) characteristics of NPN transistor 2N3904.

7.2 MOSFET Characteristics

Metal-oxide semiconductor field effect transistors (MOSFET) normally have high input resistance because of the oxide insulation between the gate and the channel. There are two types of MOSFETs: the enhancement type and the depletion type. In the enhancement type, the channel between the source and drain has to be induced by applying a voltage at the gate. In the depletion type MOSFET, the structure of the device is such that there exists a channel between the source and drain. Because the enhancement type MOSFET is widely used, the presentation of this section will be done using enhancement-type MOSFETs. The voltage needed to create the channel between the source and drain is called the threshold voltage, V_T. For n-channel enhancement MOSFETs, V_T is positive and for p-channel devices, it is negative. The electronic symbol of a MOSFET is shown in Figure 7.9.

MOSFETs can operate in three modes: cut-off, triode, and saturation regions. The following is a short description of the three regions of operation.

1. Cut-off Region

   For an n-channel MOSFET, if the gate-source voltage V_{GS} satisfies the condition

$$V_{GS} < V_T \tag{7.8}$$

   then the device is cut off. This implies that the drain current is zero for all values of the drain-to-source voltage.

2. Triode Region

   When $V_{GS} > V_T$ and V_{DS} is small, the MOSFET will be in the triode region. In the latter region, the device behaves as a nonlinear voltage-controlled resistance. The drain current I_D is related to drain source voltage V_{DS} by the expression

$$I_D = k_n\, [2(V_{GS} - V_T)V_{DS} - V_{DS}^2](1 + \lambda V_{DS}) \tag{7.9}$$

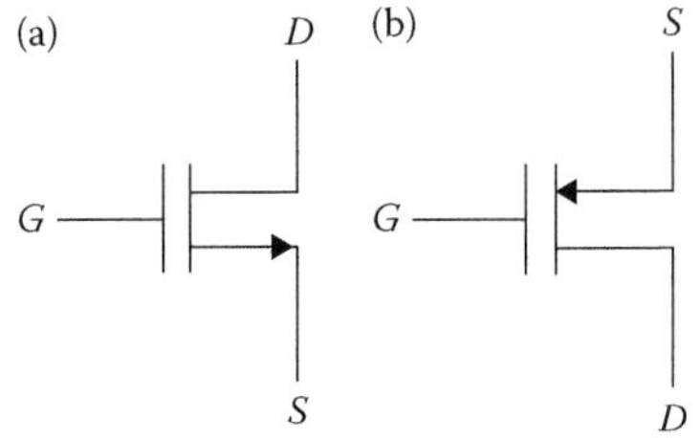

FIGURE 7.9
Circuit symbol of (a) n-channel and (b) p-channel MOSFETs.

provided

$$V_{DS} \leq V_{GS} - V_T \tag{7.10}$$

where

$$k_n = \frac{\mu_n \varepsilon \varepsilon_{OX}}{2t_{OX}} \frac{W}{L} = \frac{\mu_n C_{OX}}{2} \left(\frac{W}{L} \right) \tag{7.11}$$

and

μ_n is the surface mobility of electrons,
ε is the permittivity of free space (8.85×10^{-12} F/cm),
ε_{OX} is the dielectric constant of SiO_2,
t_{OX} is oxide thickness,
L is length of the channel,
W is width of the channel,
λ is channel width modulation factor.

3. Saturation Region
If $V_{GS} > V_T$, a MOSFET will operate in the saturation region provided,

$$V_{DS} \geq V_{GS} - V_T \tag{7.12}$$

In the saturation region, the current-voltage characteristics are given as

$$I_D = k_n (V_{GS} - V_T)^2 (1 + \lambda V_{DS}) \tag{7.13}$$

The transconductance is given as

$$g_m = \frac{\Delta I_D}{\Delta V_{GS}} \tag{7.14}$$

and the incremented drain-to-source resistance, r_{CE}, is given as

$$r_{CE} = \frac{\Delta V_{DS}}{\Delta I_{DS}} \tag{7.15}$$

In the following example, we shall obtain the I_D versus V_{DS} characteristics of a MOSFET.

Example 7.3 Output Characteristics of N-Channel MOSFET

Use the Analog Discovery board to determine the output current vs. voltage (I_D vs. V_{DS}) curve of one of the n-channel MOSFETs of the MOSFET array CD4007UB.

Solution:

Build the Circuit:

Use a breadboard to build the circuit shown in Figure 7.10. Connect the AWG1 (yellow wire) to node 3 of the circuit and the AWG2 (yellow-white wire) to node 1 of the circuit. Also, connect the scope channel 1 positive (orange wire) to node 2. In addition, connect the scope channel 1 negative (orange-white wire) to node 0. Moreover, connect the ground (black wire) to node 0 of the circuit. Furthermore, connect the scope channel 2 positive (blue wire) to node 3, and the scope channel 2 negative (blue-white wire) to node 2.

For the MOSFET array CD4007UB with pin configuration shown in Figure 7.11, connect the gate (pin #6) to node 1, drain (pin #8) to node 2, and the source (pin #7) to node 0 (ground).

Generation of the Triangular Waveform of AWG1:

AWG1 generates drain voltage, V_{DD}. The triangular waveform is made to range from 0 to 5 V. Each ramp generates an I–V curve for a specific value of the gate voltage. The triangular waveform is generated through the following commands: Click on the WaveGen in the main WaveForms screen. Go to Basic → Triangular → Frequency (set to 50 Hz) → Amplitude (set to 2.5 V) → Offset (set to 2.5 V) → Symmetry (set to 50%) → Click Run AWG1.

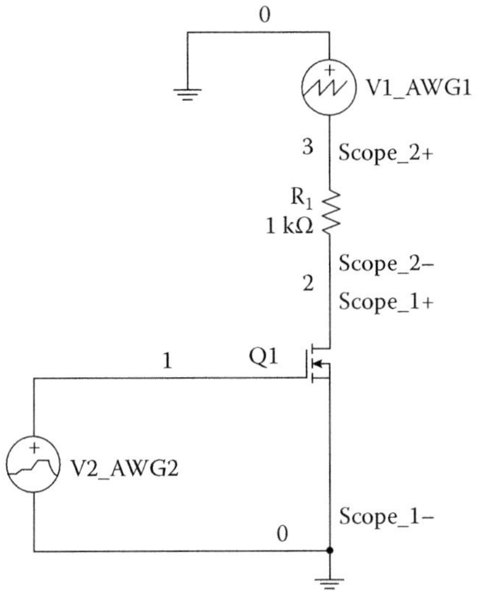

FIGURE 7.10
Circuit for determining current vs. voltage characteristics of n-channel MOSFETs.

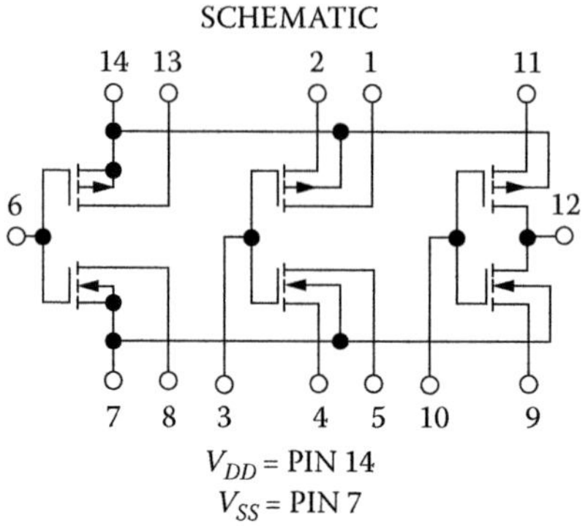

FIGURE 7.11
Pin configuration of CD4007UB MOSFET array [For the CD4007UB MOSFET array, *Pin 7* is connected to the substrate of the nMOS and *should be connected to the most negative voltage of the circuit*; pin 14 is the bulk of the PMOS and should be connected to the most positive voltage in the circuit.] (From Texas Instruments, www.ti.com, CD4007UBE datasheet.)

Generation of the Step Function of AWG2:
AWG2 generates uniformly distributed steps (stairs waveform). To generate the stairs waveform, open up an Excel file and insert the numbers 1, 2, 3, 4, 5, and 6 in six rows. The Excel file should be saved with file extension .csv or .txt.

Open the Arbitrary Waveform Generator, AWG2. Go to Custom → File → Select source file → Import data in source file with the extension .csv or .txt → Frequency (set to 10 Hz, frequency is the buffer iteration frequency, and 10 Hz means the whole stair sequence takes 100 ms) → Amplitude (set to 1.5 V) Offset (set to 2.5 V) → Symmetry (set to 50%) → Click Run AWG2. Figure 7.12 shows the waveforms generated by AWG1 and AWG2.

Activation of Scope:
Scope channel 1 measures the voltage between the drain and source (V_{DS}). Scope channel 2 measures the voltage across the resistor R2 (1000 Ω). To obtain the drain current, we calculate C2/R2. The Mathematics Channel is used to obtain the drain current. Follow the following steps to obtain the I–V curves: Add Mathematic Channel → Custom → Type C2/1000 (this current is calculated as the voltage across the 1000 Ω resistor divided by resistance 1000 Ω) → control → Add XY (this initiates the XY plot) → set X to C1 (Channel 1) and set Y to M1 (Mathematic Channel 1) → Change Math Mode Units to A → Change the M1 range to 500 μA/div. Figure 7.13 shows the I_D vs. V_{DS} characteristics of the one of the nMOS transistors of CD4007.

With regard to the output I–V curve of the MOSFET, the maximum gate voltage is 4 V. Since six steps were used in generation of the gate-source voltage, the step gate-source voltage is 0.8 V (which is 4 V divided by 5).

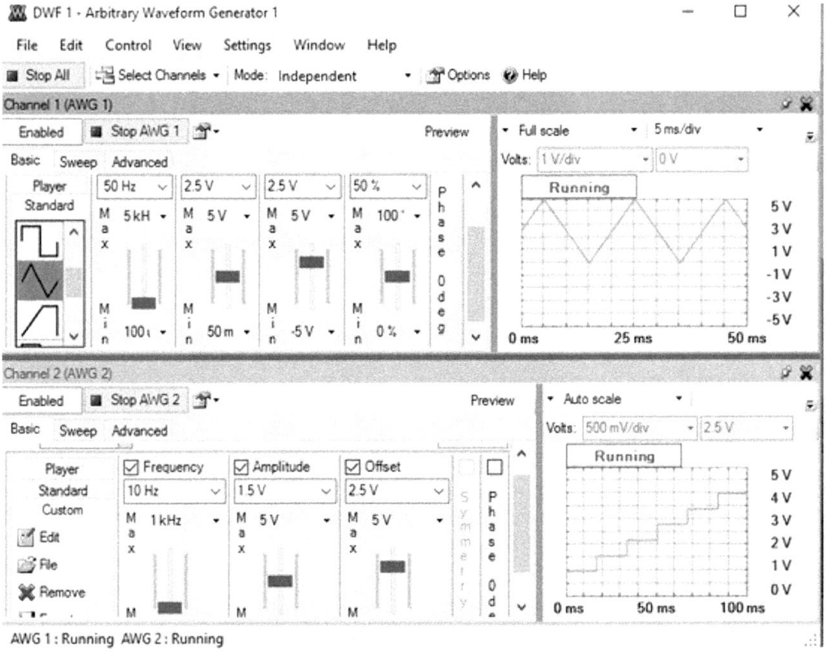

FIGURE 7.12
Arbitrary Waveform Generator 1 and 2 (AWG1 and AWG2) settings.

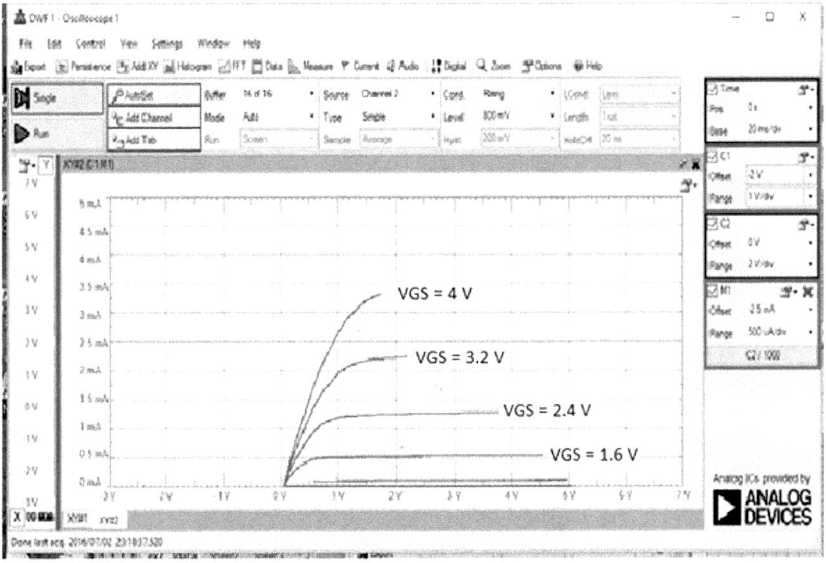

FIGURE 7.13
Output characteristics of n-channel MOSFETs.

Example 7.4 Input Characteristics of N-Channel MOSFET

Use the Analog Discovery board to determine the current (I_D) vs. voltage (V_{GS}) curve of one of the n-channel MOSFETs of the MOSFET array CD4007UB. Determine the threshold voltage of the nMOS transistor.

Solution:

When the MOSFET is in saturation region, and neglecting the channel length modulation, the MOSFET current can be given as

$$I_D = k_n (V_{GS} - V_T)^2 \tag{7.16}$$

Equation (7.16) can be rewritten as:

$$\sqrt{I_D} = \sqrt{k_n} (V_{GS} - V_T) \tag{7.17}$$

For a plot of $\sqrt{I_D}$ versus V_{GS}, the intersection with the horizontal axis gives the threshold voltage V_T, and the slope is $\sqrt{k_n}$. To force the transistor into the saturation region, the gate and the drain are connected together.

Build the Circuit:
Use a breadboard to build the circuit shown in Figure 7.14. Connect the AWG1 (yellow wire) to node 2 of the circuit. Also, connect the scope channel 1 positive (orange wire) to node 1. In addition, connect the scope channel 1 negative (orange-white wire) to node 0. Moreover, connect the ground (black wire) to node 0 of the circuit. Furthermore, connect the scope channel 2 positive (blue wire) to node 2, and the scope channel 2 negative (blue-white wire) to node 1.

For the MOSFET array CD4007UB, connect the gate (pin #6) to node 1, drain (pin #8) to node 1, and the source (pin #7) to node 0.

Generation of the Triangular Waveform of AWG1:
AWG1 generates drain voltage, which is a triangular signal. The triangular waveform is made to range from 0 to 5 V. Each ramp generates an I–V curve for a specific value drain voltage. The triangular waveform is generated through the following commands: Click on the WaveGen in the main WaveForms screen. Go to Basic → Triangular → Frequency (set to 50 Hz) → Amplitude (set to 2.5 V) → Offset (set to 2.5 V) → Symmetry (set to 50%) → Click Run AWG1. Figure 7.15 shows the setup for generating the triangular waveform from the Analog Discovery WaveGen AWG1.

Activation of Scope:
Scope channel 1 measures the voltage between the drain and source (V_{DS}). Scope channel 2 measures the voltage across the resistor R2 (1000 Ω). To obtain the drain current, we calculate C2/R2. The Mathematic Channel is used to obtain the drain current. Follow these steps to obtain the

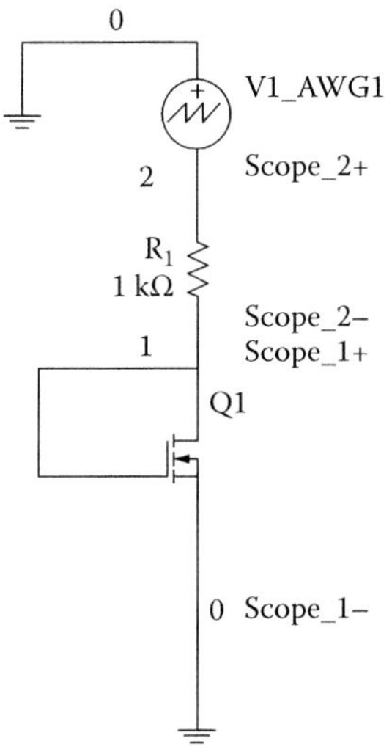

FIGURE 7.14
Circuit for determining input characteristics of n-channel MOSFETs.

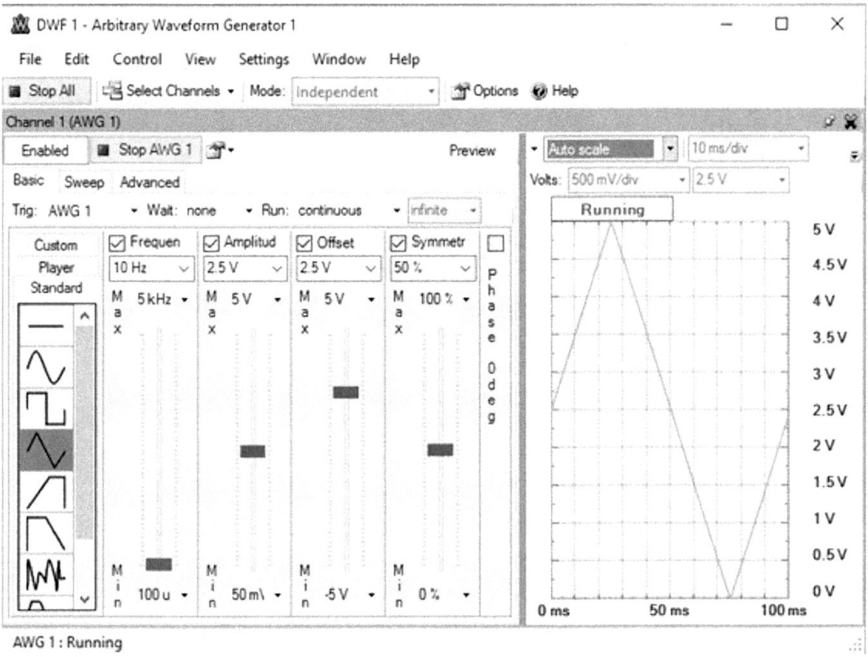

FIGURE 7.15
Arbitrary Waveform Generator 1 and 2 (AWG1) settings.

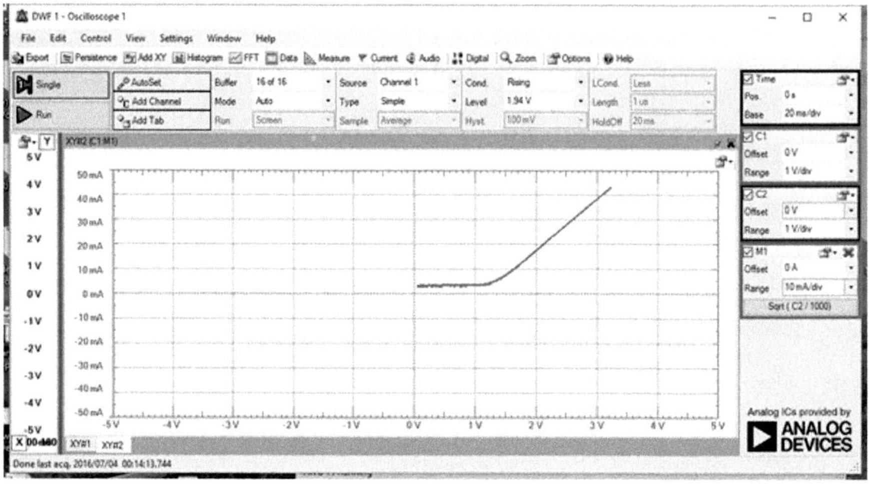

FIGURE 7.16
$\sqrt{I_D}$ versus V_{GS} characteristics of MOSFET array CD4007UB.

I–V curves: Add Mathematic Channel → Custom → Type sqrt(C2/1000) (this is the square root of the current which is calculated as the voltage across the 1000 Ω resistor divided by the resistance of 1000 Ω) → control → Add XY (this initiates the XY plot) → set X to C1 (Channel 1) and set Y to M1 (Mathematic Channel 1) → Change Math Mode Units to A → Change the M1 range to 1 mA/V. Figure 7.16 shows the input characteristics of one of the nMOS transistors CD4007.

From Equation 7.17, one can find V_T and $\sqrt{k_n}$. From Figure 7.16, the slope of the curve is $\sqrt{k_n}$ and the intersection of the curve with the horizontal axis gives the threshold voltage V_T. From Figure 7.16, the estimate of the threshold voltage is 1.2 V.

7.3 Biasing BJT Amplifiers

Biasing networks are used to establish an appropriate dc operating point for the transistor in a circuit. For stable and consistent operation, the dc operating point should be held relatively constant under varying conditions. There are several biasing circuits available in the literature. Some are for biasing discrete circuits and others for integrated circuits. Figures 7.17 and 7.18 show some biasing networks for discrete circuits.

The dc equivalent circuit for Figure 7.17 is shown in Figure 7.19. The circuit can be used to obtain the bias point of the transistor circuit.

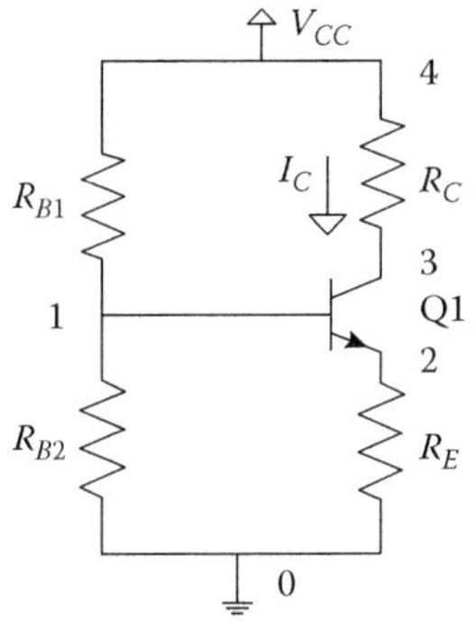

FIGURE 7.17
Biasing circuit for BJT discrete circuits with two base resistors.

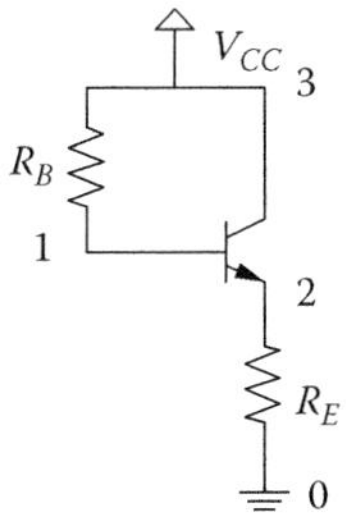

FIGURE 7.18
Biasing BJT discrete network with one base resistor.

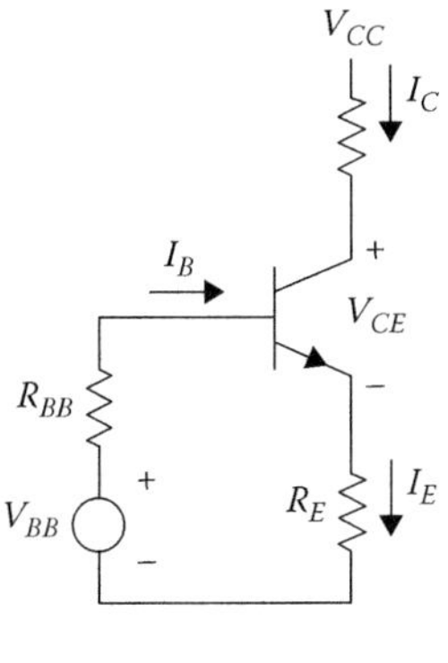

FIGURE 7.19
DC equivalent circuit of Figure 7.17.

The emitter resistor, R_E, provides stabilization of the bias point. If V_{BB} and R_B are the Thevenin equivalent parameters for the base bias circuit, then

$$V_{BB} = \frac{V_{CC}R_{B2}}{R_{B1} + R_{B2}} \tag{7.18}$$

$$R_B = R_{B1} \| R_{B2} \tag{7.19}$$

Using Kirchhoff's Voltage Law for the base circuit, we have

$$V_{BB} = I_B R_B + V_{BE} + I_E R_E \tag{7.20}$$

Using Equation 7.4 and Figure 7.19, we have

$$I_E = I_B + I_C = I_B + \beta_F I_B = (\beta_F + 1) I_B \tag{7.21}$$

Substituting Equation 7.21 into 7.20, we have

$$I_B = \frac{V_{BB} - V_{BE}}{R_B + (\beta_F + 1) R_E} \tag{7.22}$$

or

$$I_C = \frac{V_{BB} - V_{BE}}{\frac{R_B}{\beta_F} + \frac{(\beta_F + 1)}{\beta_F} R_E} \tag{7.23}$$

Applying Kirchhoff's Voltage Law at the output loop of Figure 7.19 gives

$$V_{CE} = V_{CC} - I_C R_C - I_E R_E \tag{7.24}$$

$$= V_{CC} - I_C \left(R_C + \frac{R_E}{\alpha_F} \right) \tag{7.25}$$

Equations 7.23 and 7.24 can be used to obtain the bias point of Figure 7.17. Example 7.5 shows how to obtain the bias point of a BJT amplifier.

Example 7.5 Bias Point of BJT Amplifier Circuit

Use the Analog Discovery board to determine the current (I_C) and voltage (V_{CE}) of the amplifier circuit shown in Figure 7.20. Change the collector voltage to 4.0 and 4.5 V. Determine the bias points for the previously mentioned values of collector voltage, V_{CC}.

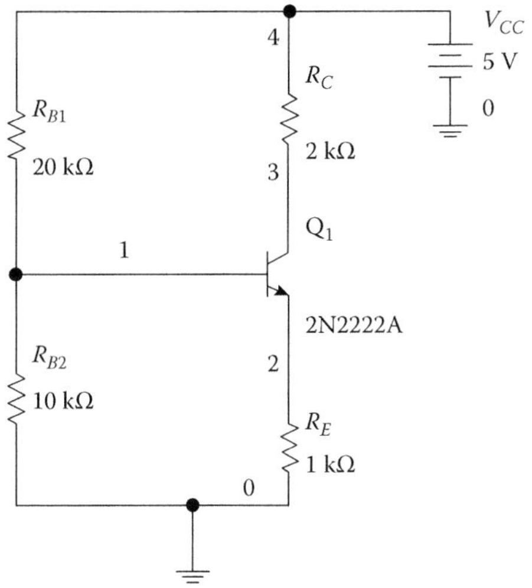

FIGURE 7.20
BJT biasing circuit.

Solution:

Build the Circuit:

Use a breadboard to build the circuit shown in Figure 7.20. Connect the AWG1 (yellow wire) to node 4 of the circuit. Also, connect the scope channel 1 positive (orange wire) to node 4. In addition, connect the scope channel 1 negative (orange-white wire) to node 3. Moreover, connect the ground (black wire) to node 0 of the circuit. Furthermore, connect the scope channel 2 positive (blue wire) to node 3, and the scope channel 2 negative (blue-white wire) to node 2.

For the transistor 2N3904, connect the base of the transistor to node 1, collector to node 3, and the emitter to node 2 of the circuit shown in Figure 7.20.

Signal Generation:

Click on the WaveGen in the main WaveForms screen. Go to Basic → dc → Offset (set to 5 V) → Click Run AWG1.

Activation of Voltmeter:

Activate the Voltmeter by clicking on Voltmeter under the More Instruments tab. Click Enable to start reading the voltage. Channel 1 measures the voltage across the collector resistor of 2.0 K. Scope channel 2 measures the voltage between the collector and the emitter (V_{CE}). To obtain the collector current, the voltage across the collector voltage is divided by a resistance of 2.0 kΩ.

Figure 7.21 shows one of the waveform generated settings when V_{CC} = 5 V. Figure 7.22 shows the voltmeter readings when the dc supply voltage was 5 V. Table 7.2 shows the readings of Channels 1 and 2 of the voltmeter for V_{CC} values of 4.0, 4.5, and 5.0 V.

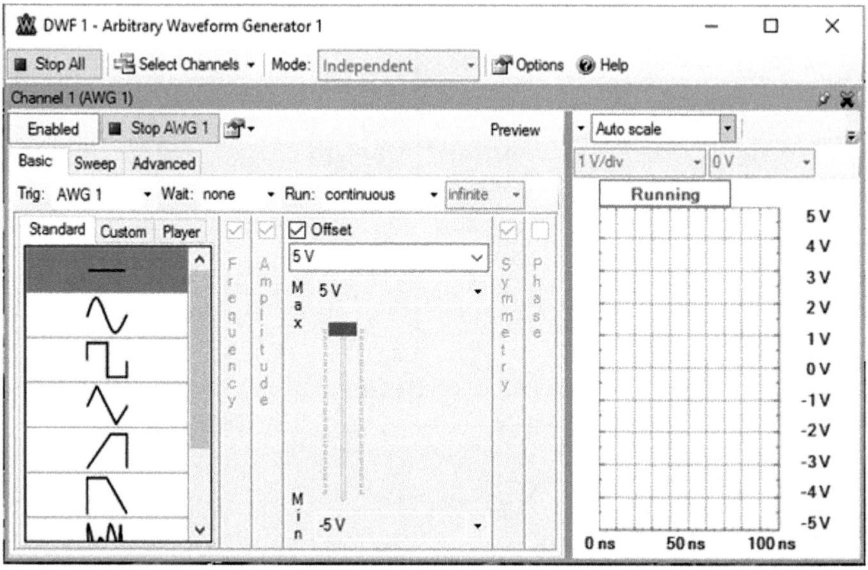

FIGURE 7.21
Arbitrary Waveform Generator for supply voltage of 5 V.

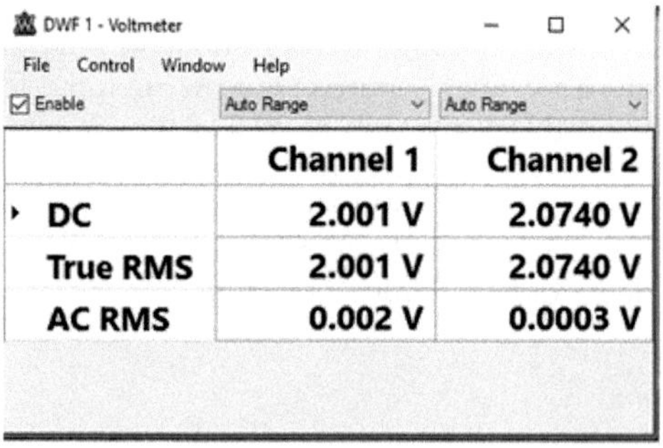

FIGURE 7.22
Voltmeter readings for V_{CC} of 5 V.

TABLE 7.2

Bias Point Measurements with Various Values of V_{CC} of Figure 7.20

Collector Voltage (V_{CC}), V	CH 1 Voltage, V	CH 2 Voltage, V	V_{CE}, V	I_C, mA
4	1.385	2.0049	2.0049	0.6925
4.5	1.698	2.0355	2.0355	0.849
5	2.001	2.074	2.074	1.0005

It can be seen from Table 7.2 that the voltage, V_{CE}, is fairly constant with respect to the changes in supply voltage V_{CC}. However, the collector current changes considerably with changes in V_{CC}.

7.4 Biasing of MOSFET Amplifiers

A popular circuit for biasing discrete MOSFET amplifiers is shown in Figure 7.23. The resistances R_{G1} and R_{G2} define the gate voltage. The resistance R_S improves operating point stability.

Because of the insulated gate, the current that passes through the gate of the MOSFET is negligible. The gate voltage is given as

$$V_G = \frac{R_{G1}}{R_{G1} + R_{G2}} V_{DD} \tag{7.26}$$

The gate-source voltage V_{GS} is

$$V_{GS} = V_G - I_S R_S \tag{7.27}$$

For conduction of the MOSFET, the gate-source voltage V_{GS} should be greater than the threshold voltage of the MOSFET, V_T. Since $I_D = I_S$, Equation 7.27 becomes

$$V_{GS} = V_G - I_D R_S \tag{7.28}$$

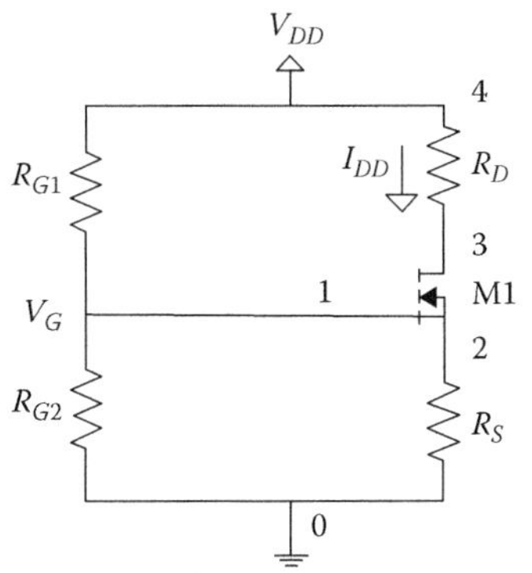

FIGURE 7.23
Biasing circuit for MOSFET using fixed gate voltage and self-bias resistors, R_S.

The drain-source voltage is obtained by using KVL at the output of the circuit

$$\begin{aligned} V_{DS} &= V_{DD} - I_D R_D - I_S R_S \\ &= V_{DD} - I_D(R_D + R_S) \end{aligned} \tag{7.29}$$

For proper operation of the bias circuit,

$$V_{GS} > V_T \tag{7.30}$$

The circuit shown in Figure 7.24 is a MOSFET transistor with the drain connected to the gate. The circuit is normally referred to as a diode-connected enhancement transistor.

From Equation 7.12, the MOSFET is in saturation provided

$$V_{DS} > V_{GS} - V_T$$

That is

$$V_{DS} - V_{GS} > -V_T \text{ or } V_{DS} + V_{SG} > -V_T$$

or

$$V_{DG} > -V_T \tag{7.31}$$

Since $V_{DG} = 0$ and V_T is positive for an n-channel MOSFET, the device of Figure 7.24 is in saturation. From Equation 7.16

$$i_D = k_n\left(V_{GS} - V_T\right)^2$$

But if $V_{GS} = V_{DS}$, Equation 7.16 becomes

$$i_D = k_n\left(V_{DS} - V_T\right)^2 \tag{7.32}$$

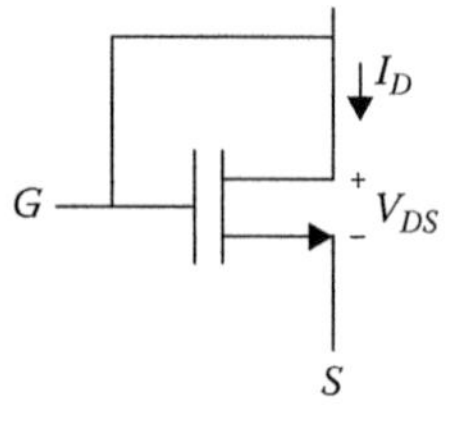

FIGURE 7.24
Diode-connected enhancement type MOSFET.

The diode-connected enhancement MOSFET can also be used to generate dc currents for nMOS and CMOS analog integrated circuits. Example 7.6 shows how to obtain the bias point of a MOSFET amplifier.

Example 7.6 Bias Point of a MOSFET Amplifier Circuit

Use the Analog Discovery board to determine the current (I_D) and voltage (V_{DS}) of the amplifier circuit shown in Figure 7.25. Change the drain supply voltage to 4.0 and 4.5 V. Determine the bias points for the previously mentioned values of supply voltage, V_{DD}.

Solution:

Build the Circuit:

Use a breadboard to build the circuit shown in Figure 7.25. Connect the AWG1 (yellow wire) to node 4 of the circuit. Also, connect the scope channel 1 positive (orange wire) to node 4. In addition, connect the scope channel 1 negative (orange-white wire) to node 3. Moreover, connect the scope channel 2 positive (blue wire) to node 3, and the scope channel 2 negative (blue-white wire) to node 2. Furthermore, connect the ground (black wire) to node 0 of the circuit.

For the transistor BS170, connect the gate of the transistor to node 1, the drain to node 3, and the source to node 2 of the circuit.

Signal Generation:

Click on the WaveGen in the main WaveForms screen. Go to Basic → dc → Offset (set to 5 V) → Click Run AWG1. Figure 7.26 shows the setup of Arbitrary Waveform Generator, AWG1, when V_{DD} = 5 V.

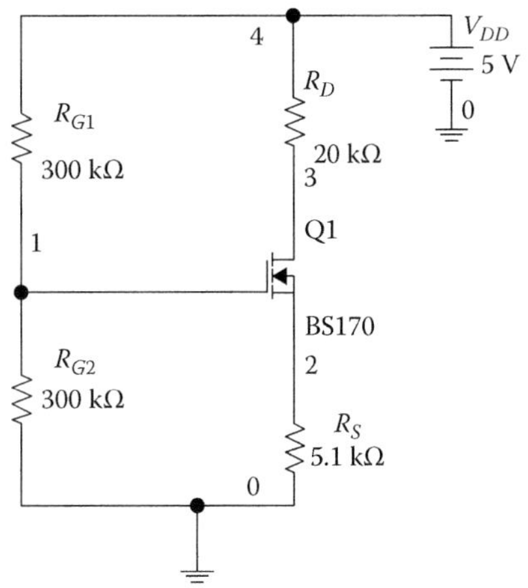

FIGURE 7.25
MOSFET biasing circuit.

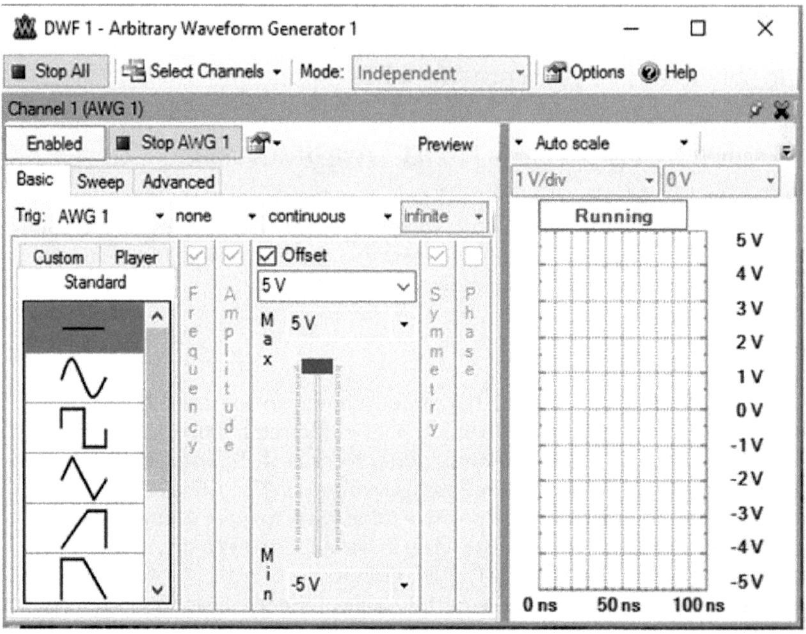

FIGURE 7.26
Arbitrary Waveform Generator for supply voltage of 5 V.

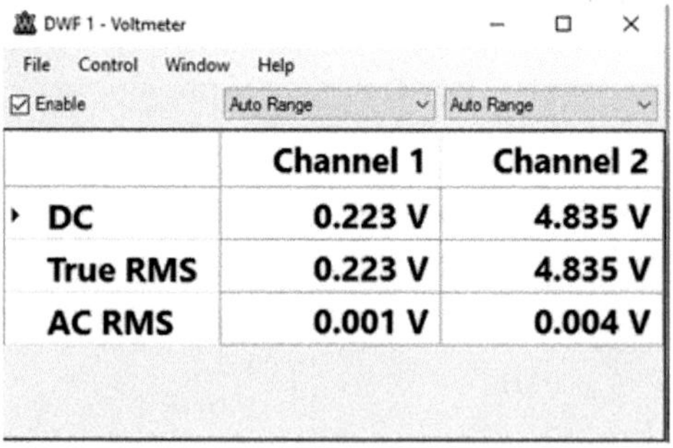

FIGURE 7.27
Voltmeter readings for V_{DD} of 5 V.

Activation of Voltmeter:

Activate the Voltmeter by clicking Voltmeter under the More Instruments tab. Click Enable to start reading the voltage. Channel 1 measures the voltage across the drain resistor, 20 kΩ resistor. Scope channel 2 measures the voltage between the drain and the source (V_{DS}). To obtain the drain current, the voltage across the collector voltage is divided by 20 kΩ. Figure 7.27 shows the voltmeter readings when the dc supply voltage was 5 V.

TABLE 7.3

Bias Point Measurements with Various Values of V_{DD} of Figure 7.25

Drain Voltage (V_{DD}), V	CH 1 Voltage, V	CH 2 Voltage, V	V_{DS}, V	I_D, mA
4	0.187	3.872	3.872	0.00935
4.5	0.206	4.354	4.354	0.0103
5	0.223	4.835	4.835	0.01115

Table 7.3 shows the readings of Channels 1 and 2 of the voltmeter for V_{DD} values of 4.0, 4.5, and 5.0 V. Channel 1 measured the voltage across the drain resistance, R_D, of Figure 7.25. The drain current is obtained by dividing the Channel 1 voltage by the drain resistance of 20 kΩ. Channel 2 measured the voltage between the drain and source of transistor BS170 of Figure 7.25. It can be seen from Table 7.3 that the voltage, V_{DS}, decreases with respect to the changes in supply voltage V_{DD}. In addition, the drain current also decreases with changes in V_{DD}.

7.5 Frequency Response of BJT Amplifiers

Amplifiers are normally used for voltage amplification, current amplification, impedance matching or to provide isolation between amplifier stages. Transistor amplifiers can be built using bipolar junction transistors. Amplifiers built using BJT can be common-emitter, common-collector (emitter follower), or common-base amplifiers. Common-emitter amplifiers have relatively high voltage gain. A common-emitter amplifier is shown in Figure 7.28. The amplifier is capable of generating a relatively high current

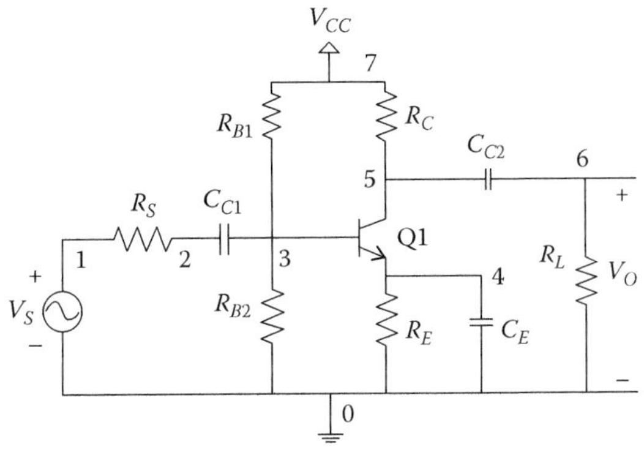

FIGURE 7.28
Common-emitter amplifier.

and high voltage gains. The input resistance is medium and is essentially independent of the load resistance, R_L.

For the common-emitter amplifier, shown in Figure 7.28, the coupling capacitor, C_{C1}, couples the voltage source, V_S, to the bias network. Coupling capacitor, C_{C2}, connects the collector resistance, R_C, to the load R_L. The bypass capacitance, C_E, is used to increase the midband gain, since it effectively short circuits the emitter resistance R_E at midband frequencies. The resistance R_E is needed for bias stability. The external capacitors C_{C1}, C_{C2}, C_E influence the low frequency response of the common-emitter amplifier. The internal capacitances of the transistor control the high frequency cut-off. The overall gain of the common-emitter amplifier can be written as

$$A(s) = \frac{A_m s^2 (s + w_z)}{(s + w_{L1})(s + w_{L2})(s + w_{L3})(1 + s/w_H)} \tag{7.33}$$

where

A_M is the midband gain,

w_H is the frequency of the dominant high frequency pole,

w_{L1}, w_{L2}, w_{L3} are low frequency poles introduced by the coupling and bypass capacitors,

w_Z is the zero introduced by the bypass capacitor.

The midband gain is obtained by short-circuiting all the external capacitors and open-circuiting the internal capacitors. Figure 7.29 shows the equivalent for calculating the midband gain.

From Figure 7.29, the midband gain, A_m, is

$$A_m = \frac{V_O}{V_S} = -\beta\left[r_{CE} \| R_C \| R_L\right]\left[\frac{R_B}{R_B + r_\pi}\right]\left[\frac{1}{R_S + \left[R_B \| r_\pi\right]}\right] \tag{7.34}$$

It can be shown that the low frequency poles, w_{L1}, w_{L2}, w_{L3}, can be obtained by the following equations

$$\tau_1 = \frac{1}{w_{L1}} = C_{C1} R_{IN} \tag{7.35}$$

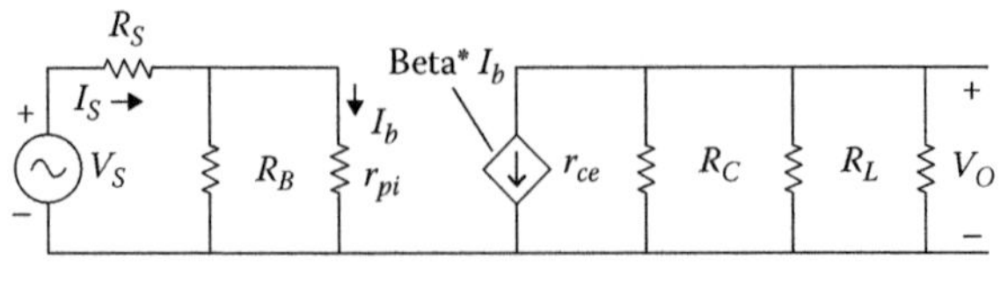

FIGURE 7.29
Equivalent circuit for calculating midband gain.

where

$$R_{IN} = R_S + \left[R_B \| r_\pi \right] \tag{7.36}$$

$$\tau_2 = \frac{1}{w_{L2}} = C_{C2}\left[R_L + \left(R_C \| r_{ce} \right) \right] \tag{7.37}$$

and

$$\tau_3 = \frac{1}{w_{L3}} = C_E R_E' \tag{7.38}$$

where

$$R_E' = R_E \left\| \left[\frac{r_\pi}{\beta_F + 1} + \left(\frac{R_B \| R_S}{\beta_F + 1} \right) \right] \right. \tag{7.39}$$

and the zero

$$w_Z = \frac{1}{R_E C_E} \tag{7.40}$$

Normally, $w_Z < w_{L3}$ and the low frequency cut-off w_L is larger than the largest pole frequency. The low frequency cut-off can be approximated as

$$w_L \cong \sqrt{(w_{L1})^2 + (w_{L2})^2 + (w_{L3})^2} \tag{7.41}$$

Example 7.7 and Example 7.8 explore the characteristics BJT amplifiers.

Example 7.7 Frequency Response of Common-Emitter Amplifier

Use the Analog Discovery board to determine (1) the magnitude response of the circuit shown in Figure 7.30, (2) midband gain, (3) low cut-off frequency, (4) high cut-off frequency, and (5) bandwidth of the amplifier.

Solution:

Build the Circuit:

Use a breadboard to build the circuit shown in Figure 7.30. Connect the AWG1 (yellow wire) to node 7 of the circuit. Also, connect the scope channel 1 positive (orange wire) to node 7 and connect the scope channel 1 negative (orange-white wire) to node 0. In addition, connect the scope channel 2 positive (blue wire) to node 6, and the scope channel 2 negative (blue-white wire) to node 0. Furthermore, connect the ground

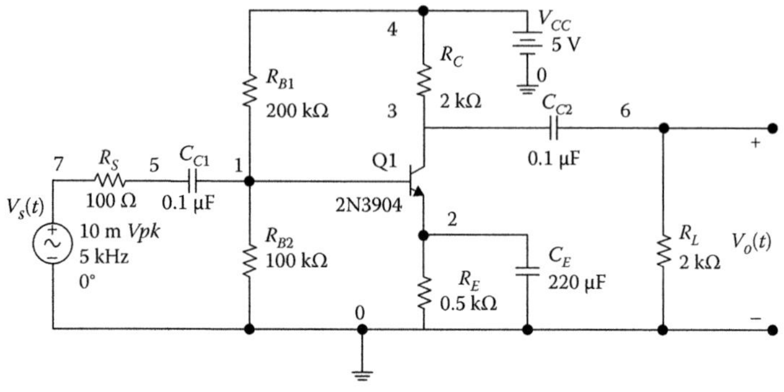

FIGURE 7.30
Common-emitter amplifier circuit.

(black wire) to node 0 of the circuit. For the transistor 2N3904, connect the base of the transistor to node 1, the collector to node 3, and the emitter to node 2 of Figure 7.30. Moreover, connect +5 V dc power supply (red wire) to node 4.

Generation of AC Sine Wave:
Click on the WaveGen in the main WaveForms screen. Go to Basic → Sine → Frequency (set to 5000 Hz) → Amplitude (set to 10.0 mV) → Offset (set to 0 V) → Symmetry (set to 50%) → Click Run AWG1. The setup of the Arbitrary Waveform Generator AWG1 is shown in Figure 7.31.

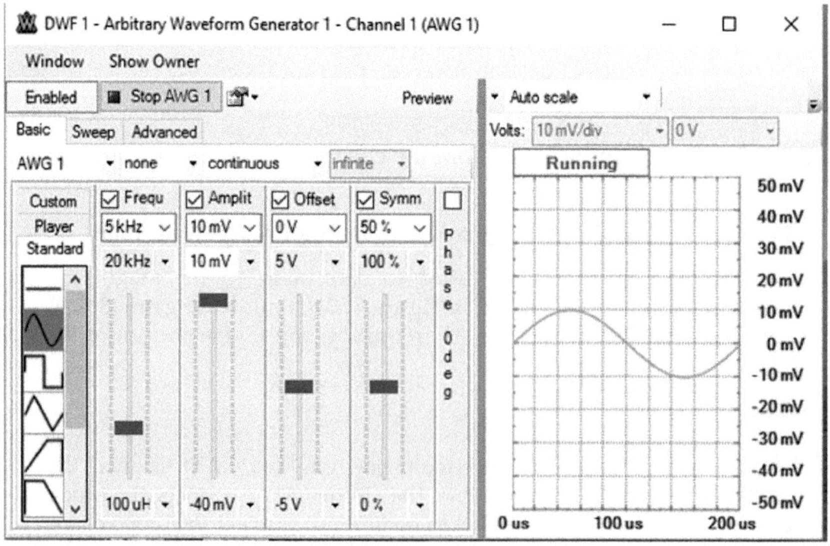

FIGURE 7.31
Arbitrary Waveform Generator for sinusoidal signal.

Activation of the Power Supply:
Click Voltage in the main WaveForms screen → Click Power to turn on the voltage sources → Click V+ to turn on the 5 V source → Upon completion of the measurements, click Power again to turn it off.

Activation of Network Analyzer:
The Network Analyzer uses both oscilloscope channels as the input and output channels. Channel 1 is for measuring the input signal and channel 2 is for measuring the output signal. Open up the Network Analyzer. Use the following setup values:

Start Frequency: 50 Hz
Stop Frequency: 10 MHz
Offset: 0 V
Input Signal Amplitude: 10 mV
Max Filter Gain: 1X
Bode Scale: Magnitude—Top 40 dB, Range of 50 dB
Phase—Top 45°, Range of 360°
Scope Channel Gain: Channel 1: 1X; Channel 2:1X
Obtain a single sweep in frequency by clicking Single. This provides a Bode plot representation of the frequency response of the circuit. The Bode plot obtained is shown in Figure 7.32.

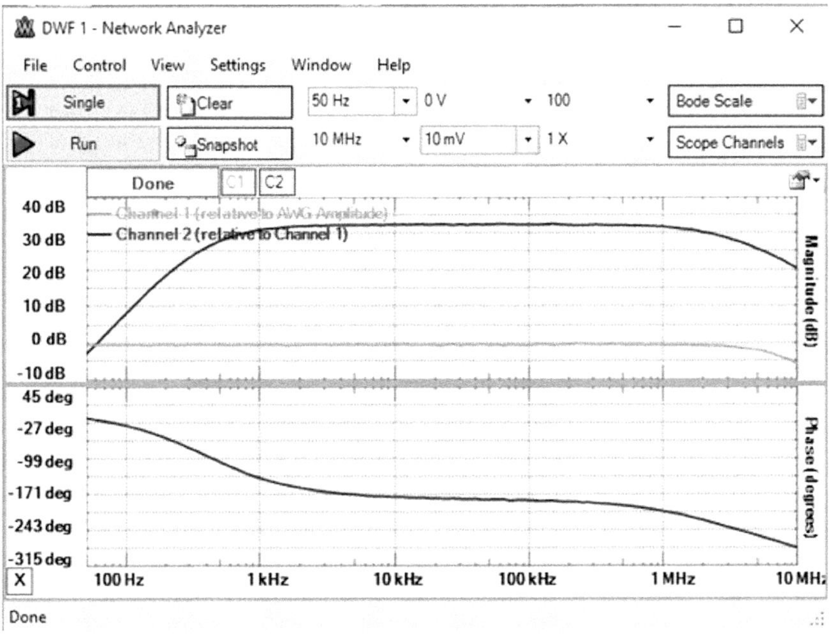

FIGURE 7.32
Magnitude and phase response of the amplifier of Figure 7.30.

From Figure 7.32, the following measurements were obtained:

Low cut-off frequency = 674.82 Hz

High cut-off frequency = 2.63 MHz

Bandwidth = 2.63 MHz – 674.82 Hz = 2.629 MHz

Midband Gain = 32.5 dB

Example 7.8 Characteristics of Emitter Follower Circuit

Use the Analog Discovery board (1) to determine gain at 5 kHz and (2) to measure the voltage across the 100 Ω resistor and calculate the input resistance.

Solution:

Build the Circuit:

Use a breadboard to build the circuit shown in Figure 7.33. Connect the AWG1 (yellow wire) to node 5 of the circuit. Also, connect the scope channel 1 positive (orange wire) to node 5. In addition, connect the scope channel 1 negative (orange-white wire) to node 0. Furthermore, connect the scope channel 2 positive (blue wire) to node 3, and the scope channel 2 negative (blue-white wire) to node 0. Moreover, connect the ground (black wire) to node 0 of the circuit. For the transistor 2N3904, connect the base of the transistor to node 1, the collector to node 2, and the emitter to node 3 of Figure 7.33. In addition, connect +5 V dc power supply (red wire) to node 2.

Generation of AC Sine Wave:

Click on the WaveGen in the main WaveForms screen. Go to Basic → Sine → Frequency (set to 1000 Hz) → Amplitude (set to 20.0 mV) → Offset (set to 0 V) → Symmetry (set to 50%) → Click Run AWG1. The setup of the Arbitrary Waveform Generator AWG1 is shown in Figure 7.34.

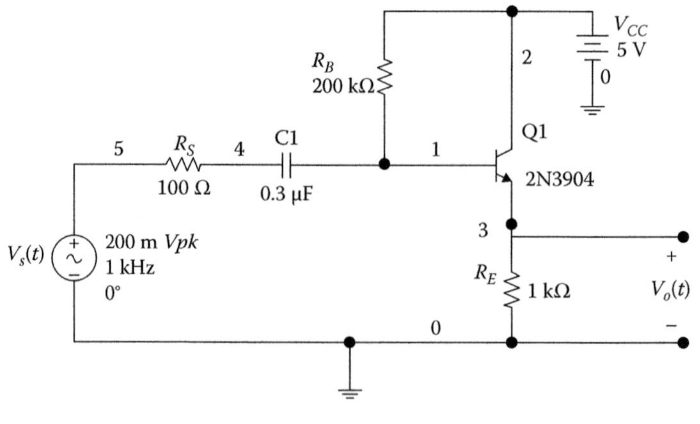

FIGURE 7.33
Emitter follower circuit.

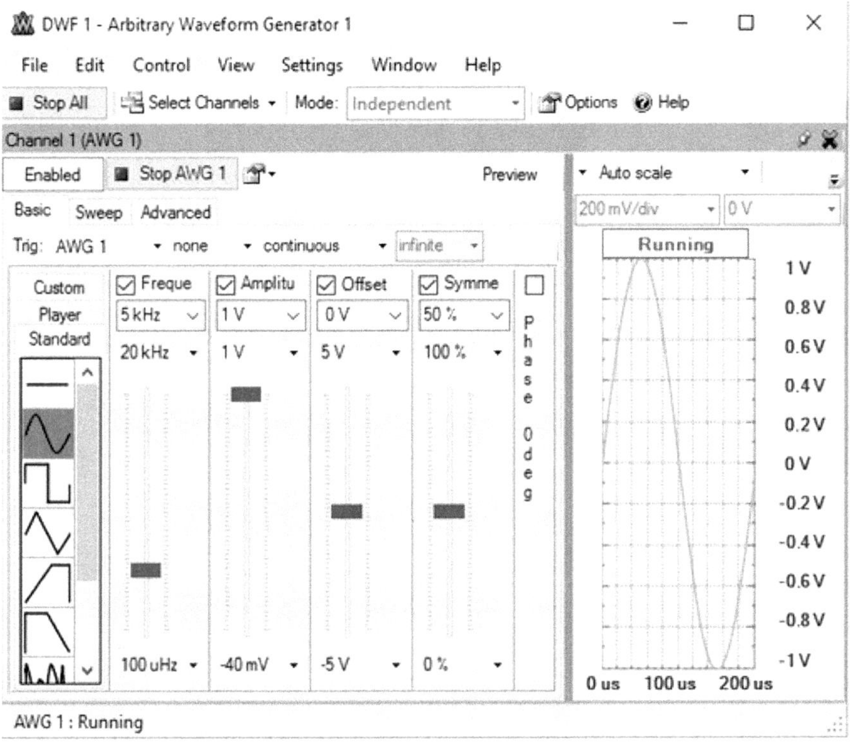

FIGURE 7.34
Arbitrary Waveform Generator for sinusoidal signal.

> *Activation of the Power Supply*
> Click Voltage in the main WaveForms screen → Click Power to turn on the voltage sources → Click V+ to turn on the 5 V source → Upon completion of the measurements, click Power again to turn it off.
>
> *Activation of Scope:*
> Click on the Scope in the main WaveForms screen. Go to Run → Autoset → Measure → Add → Channel 1 → Horizontal → Frequency → Add Selected Measurement → Add → Channel 1 → Vertical → Peak to Peak → Add Selected Instrument → Add → Channel 2 → Vertical → Peak to Peak → Add Selected Measurement → Close Add Instrument. Figure 7.35 shows the input and output waveforms displayed on the scope.

From Figure 7.35, the gain of the amplifier is

$$\text{Gain} = 396.8\ \text{mV}/403.2\ \text{mV} = 0.984$$

With the input changed to 2 V peak, the input voltage and the voltage across the resistor R_S are shown in Figure 7.36.

Input Voltage = VS = 4.05 V peak-to-peak

Voltage across Resistor R_S = 10 mV peak-to-peak

Value of Resistor RS = 100 Ω

Current flowing through RS = I = 10 mV/100 = 0.1 mA

Input Resistance = VS/I = (4.05 V/0.1 mA) = 40.5 kΩ

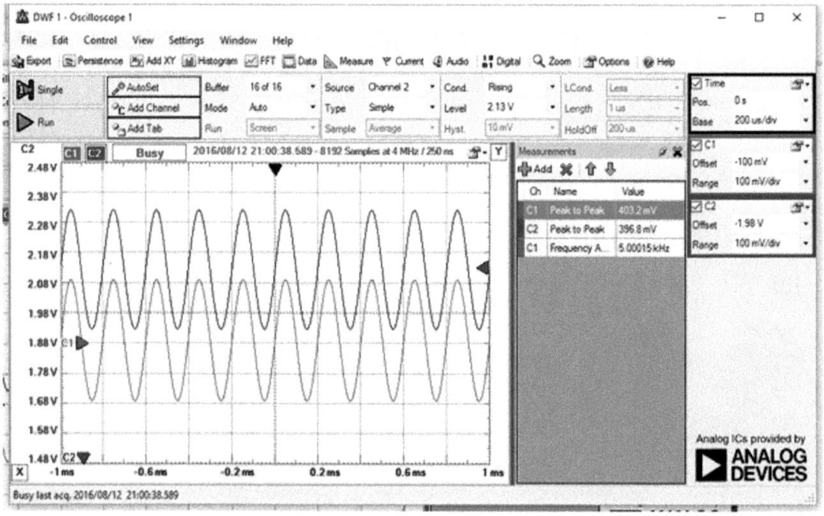

FIGURE 7.35
Display of input and output sinusoidal waveforms of the emitter follower.

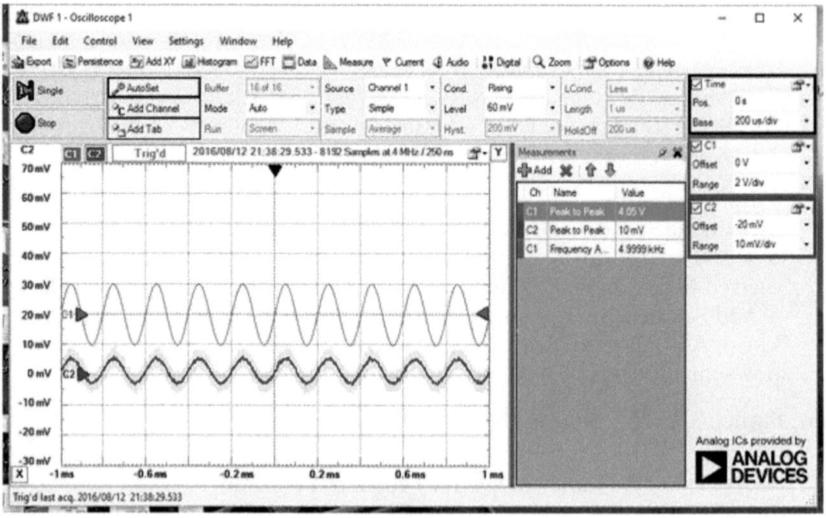

FIGURE 7.36
Waveforms for obtaining the input resistance of the emitter follower.

7.6 Frequency Response of MOSFET Amplifiers

MOSFET amplifiers can be common-source, common-drain, or common-drain common-gate amplifiers. A common-drain amplifier has relatively high input resistance, low output resistance with voltage gain that is almost equal to unity. The common-source amplifier, shown in Figure 7.37, has characteristics similar to those of the common-emitter amplifier. However, the common-source amplifier has a higher input resistance than that of the common-emitter amplifier.

The external capacitors C_{C1}, C_{C2}, and $C_{S,}$ will influence the low frequency response. The internal capacitances of the FET will affect the high frequency response. The midband gain, A_m, is obtained from the midband equivalent circuit of the common-source amplifier, shown in Figure 7.38. The equivalent circuit is obtained by short-circuiting all the external capacitors and open-circuiting all the internal capacitances of the FET.

Using voltage division,

$$v_{gs} = \frac{R_G}{R_I + R_G} v_S \tag{7.42}$$

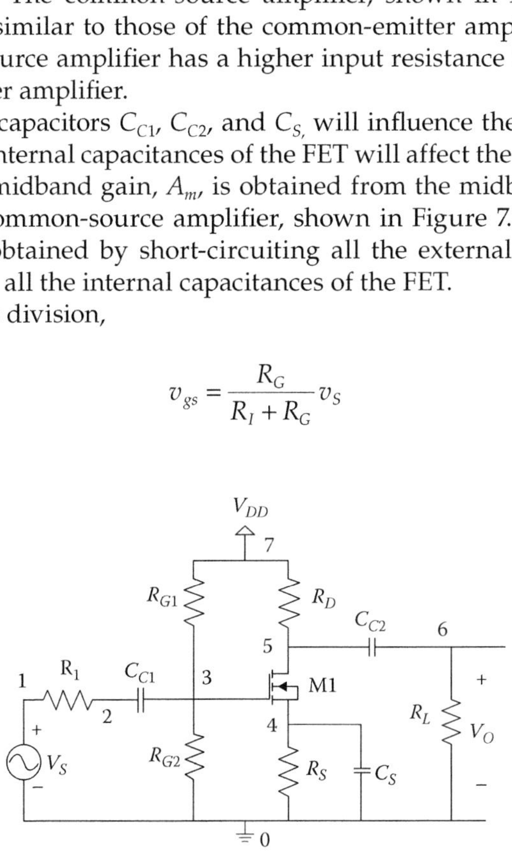

FIGURE 7.37
Common-source amplifier.

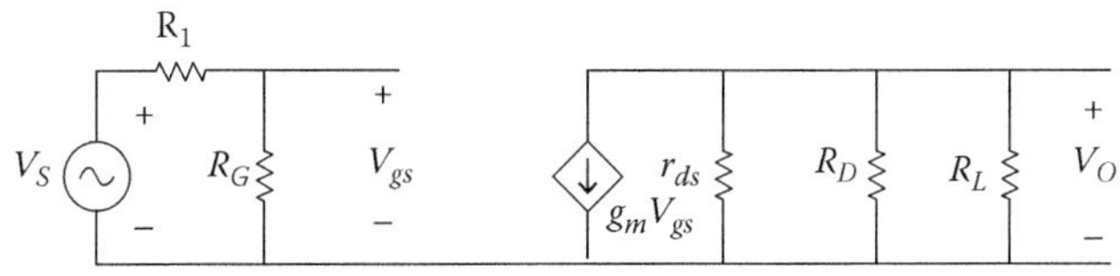

FIGURE 7.38
Midband equivalent circuit of common-source amplifier.

From Ohm's Law,

$$v_0 = -g_m v_{gs}\left(r_{ds}\left\|R_D\right\|R_L\right) \tag{7.43}$$

Substituting Equation 7.42 into 7.43, we obtain the midband gain as

$$A_m = \frac{v_0}{v_s} = -g_m\left(\frac{R_G}{R_G + R_I}\right)\left(r_{ds}\left\|R_D\right\|R_L\right) \tag{7.44}$$

At low frequencies, the small signal equivalent circuit of the common-source amplifier is shown in Figure 7.39.

It can be shown that the low frequency poles due to C_{C1} and C_{C2} can be written as

$$\tau_1 = \frac{1}{w_{L1}} \cong C_{C1}(R_g + R_I) \tag{7.45}$$

$$\tau_2 = \frac{1}{w_{L2}} \cong C_{C2}(R_L + R_D \| r_{ds}) \tag{7.46}$$

Assuming r_d is very large, the pole due to the bypass capacitance C_S can be shown to be

$$\tau_3 = \frac{1}{w_{L3}} \cong C_S\left(\frac{R_S}{1 + g_m R_S}\right) \tag{7.47}$$

and the zero of C_S is

$$w_Z = \frac{1}{R_S C_S} \tag{7.48}$$

The 3-dB frequency at the low frequency can be approximated as

$$w_L \cong \sqrt{(w_{L1})^2 + (w_{L2})^2 + (w_{L3})^2} \tag{7.49}$$

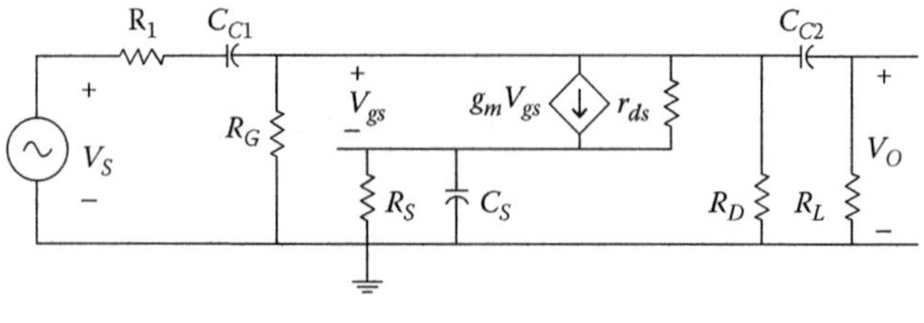

FIGURE 7.39
Equivalent circuit for obtaining the poles at low frequencies of common-source amplifier.

For a single-stage common-source amplifier, the source bypass capacitor is usually the determining factor in establishing the low 3-dB frequency. In Example 7.9 and Example 7.10, the characteristics of MOSFET amplifiers are explored.

Example 7.9 Frequency Response of Common Source Amplifier

Use the Analog Discovery board to determine (1) the magnitude response of the circuit shown in Figure 7.40, (2) midband gain, (3) low cut-off frequency, (4) high cut-off frequency, and (5) bandwidth of the amplifier.

Solution:

Build the Circuit:

Use a breadboard to build the circuit shown in Figure 7.40. Connect the AWG1 (yellow wire) to node 7 of the circuit. Also, connect the scope channel 1 positive (orange wire) to node 7. In addition, connect the scope channel 1 negative (orange-white wire) to node 0. Furthermore, connect the scope channel 2 positive (blue wire) to node 6, and the scope channel 2 negative (blue-white wire) to node 0. Moreover, connect the ground (black wire) to node 0 of the circuit. For the transistor BS170, connect the gate of the transistor to node 1, the drain to node 3, and the source to node 2. In addition, connect +5 V dc power supply (red wire) to node 4.

Generation of AC Sine Wave:

Click on the WaveGen in the main WaveForms screen. Go to Basic → Sine → Frequency (set to 5000 Hz) → Amplitude (set to 50.0 mV) → Offset (set to 0 V) → Symmetry (set to 50%) → Click Run AWG1. The setup of the Arbitrary Waveform Generator AWG1 is shown in Figure 7.41.

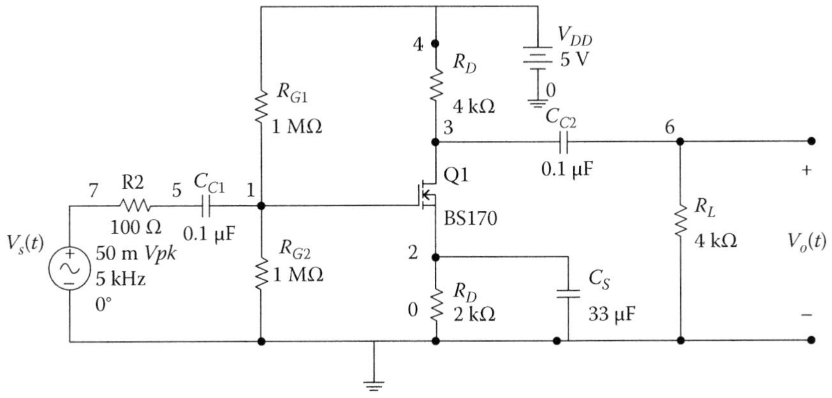

FIGURE 7.40
Common-source amplifier circuit.

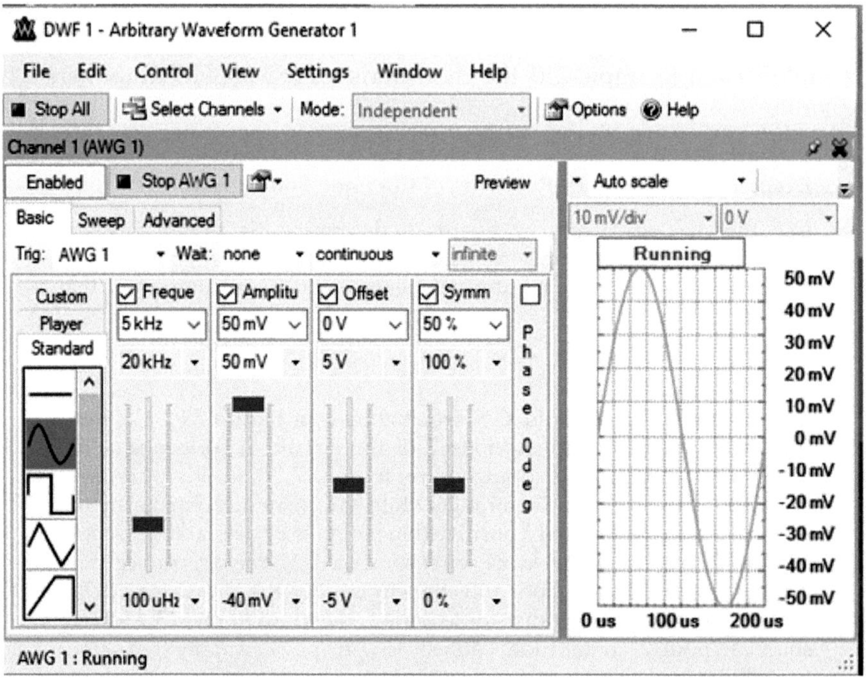

FIGURE 7.41
The setup of Arbitrary Waveform Generator AWG1.

Activation of the Power Supply:
Click Voltage in the main WaveForms screen → Click Power to turn on the voltage sources → Click V+ to turn on the 5 V source → Upon completion of the measurements, click Power again to turn it off.

Activation of Network Analyzer:
The Network Analyzer uses both oscilloscope channels as the input and output channels. Channel 1 is for measuring the input signal and Channel 2 is for measuring the output signal.

Open up the Network Analyzer. Use the following setup values:

Start Frequency: 20 Hz
Stop Frequency: 10 MHz
Offset: 0 V
Input Signal Amplitude: 50 mV
Max Filter Gain: 1X
Bode Scale: Magnitude—Top 10 dB, Range of 40 dB
Phase—Top 0°, Range of 360°
Scope Channel Gain: Channel 1: 1X; Channel 2:1X
Obtain a single sweep in frequency by clicking Single. This provides a Bode plot representation of the frequency response of the circuit. The Bode plot obtained is shown in Figure 7.42.

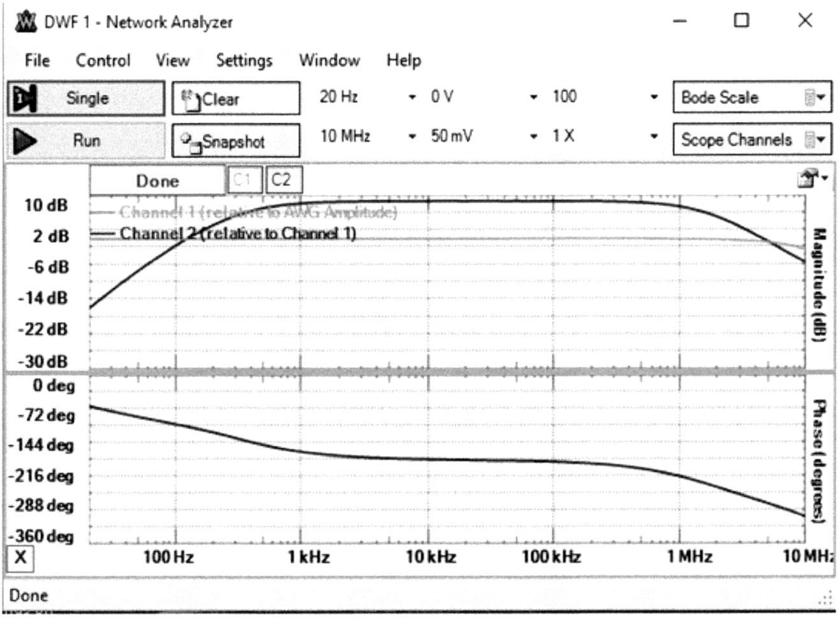

FIGURE 7.42
Magnitude and phase characteristics of the amplifier shown in Figure 7.40.

From Figure 7.42, the following measurements were obtained:

Low Cut-off Frequency = 328.17 Hz
High Cut-off Frequency = 1.77 MHz
Bandwidth = 1.77 MHz – 328.17 Hz = 1.7698 MHz
Midband Gain = 8.72 dB

Example 7.10 Frequency Response of Common-Drain Amplifier

Use the Analog Discovery board to determine (1) the magnitude response of the circuit shown in Figure 7.43, (2) midband gain, and (3) low cut-off frequency.

Solution:
Build the Circuit:
Use a breadboard to build the circuit shown in Figure 7.43. Connect the AWG1 (yellow wire) to node 7 of the circuit. Also, connect the scope channel 1 positive (orange wire) to node 7. In addition, connect the scope channel 1 negative (orange-white wire) to node 0. Furthermore, connect the scope channel 2 positive (blue wire) to node 6, and the scope channel 2 negative (blue-white wire) to node 0. Moreover, connect the ground (black wire) to node 0 of the circuit. For the transistor 2N7000, connect the gate of the transistor to node 1, drain to node 4, and the

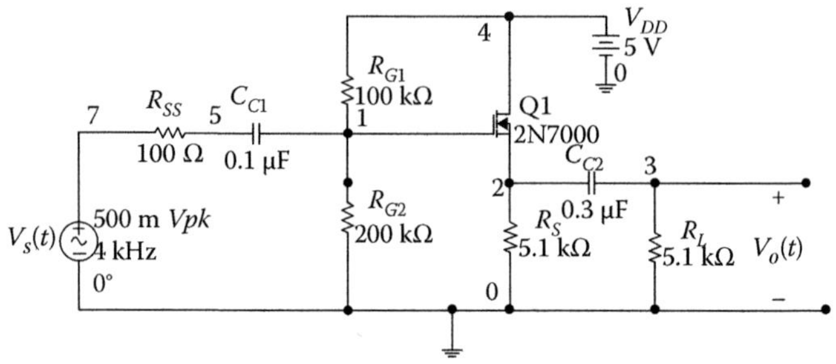

FIGURE 7.43
MOSFET amplifier.

source to node 2 of Figure 7.43. In addition, connect the +5 V dc power supply (red wire) to node 4.

Generation of AC Sine Wave:
Click on the WaveGen in the main WaveForms screen. Go to Basic → Sine → Frequency (set to 4000 Hz) → Amplitude (set to 500.0 mV) → Offset (set to 0 V) → Symmetry (set to 50%) → Click Run AWG1. The setup of the Arbitrary Waveform Generator AWG1 is shown in Figure 7.44.

Activation of the Power Supply:
Click Voltage in the main WaveForms screen → Click Power to turn on the voltage sources → Click V+ to turn on the 5 V source → Upon completion of the measurements, click Power again to turn it off.

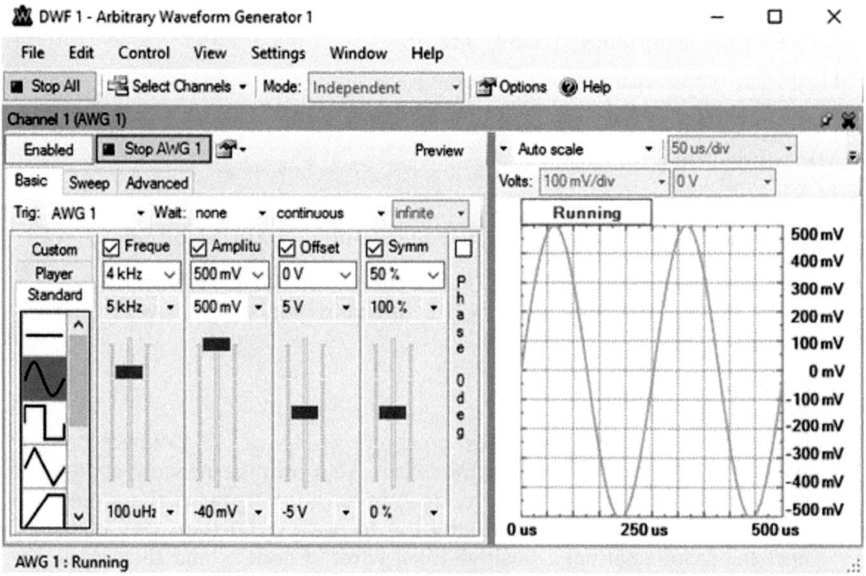

FIGURE 7.44
The setup of the Arbitrary Waveform Generator AWG1.

Activation of Network Analyzer:
The Network Analyzer uses both oscilloscope channels as the input and output channels. Channel 1 is for measuring the input signal and Channel 2 is for measuring the output signal.

Open up the Network Analyzer. Use the following setup values:

Start Frequency: 50 Hz
Stop Frequency: 10 MHz
Offset: 0 V
Input Signal Amplitude: 50 mV
Max Filter Gain: 1X
Bode Scale: Magnitude—Top 0 dB, Range of 40 dB
Phase—Top –165°, Range of 360°
Scope Channel Gain: Channel 1: 1X; Channel 2:1X
Obtain a single sweep in frequency by clicking *Single*. This provides a Bode plot representation of the frequency response of the circuit. The Bode plot obtained is shown in Figure 7.45.

From Figure 7.45, the following measurements were obtained:

Low Cut-off Frequency = 133.75 Hz

Midband Gain = –0.76 dB

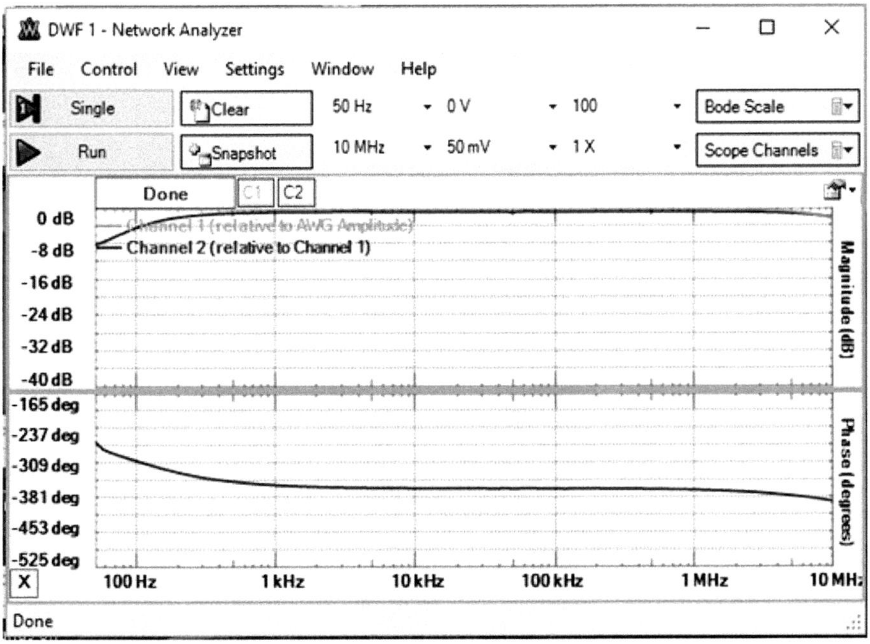

FIGURE 7.45
Bode plot of Example 7.10.

PROBLEMS

Problem 7.1 Find the output characteristics (I_C versus V_{CE}) of the NPN transistor 2N2222. You can use the circuit shown in Figure P7.1 to obtain the I–V curve.

Problem 7.2 Find the input characteristics (I_B versus V_{BE}) of the NPN transistor 2N2222. You can use the circuit similar to that shown in Figure 7.6 to obtain the I–V curve.

Problem 7.3 Find the output characteristics (I_D versus V_{DS}) of the n-channel MOSFET BS170. You can use the circuit similar to Figure 7.10 to obtain the I–V curve.

Problem 7.4 Use the Analog Discovery board to determine the current (I_D) vs. voltage (V_{GS}) curve of the n-channel MOSFET BS170. Determine the threshold voltage of the nMOS transistor.

Problem 7.5 Use the Analog Discovery board to determine the current (I_C) and voltage (V_{CE}) of the amplifier circuit shown in Figure P7.5. Change the collector voltage to 3.5, 4.0 and 4.5 V and determine the bias points for the values of collector voltage, V_{CC}.

Problem 7.6 Use the Analog Discovery board to determine the current (I_C) and voltage (V_{EC}) of the amplifier circuit shown in Figure P7.6.

Problem 7.7 Use the Analog Discovery board to determine the drain current (I_D) and voltage (V_{DS}) of the amplifier circuit shown in Figure P7.7. What will be the value of the drain current and voltage, V_{DS}, if the source resistance R_S is reduced to 500 Ω?

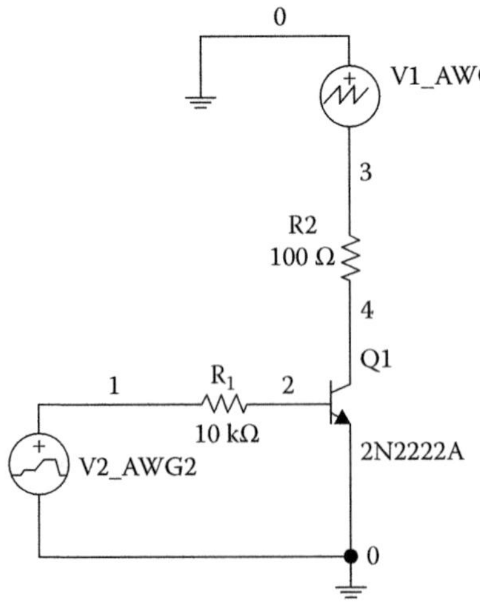

FIGURE P7.1
Circuit for output characteristic of an NPN transistor.

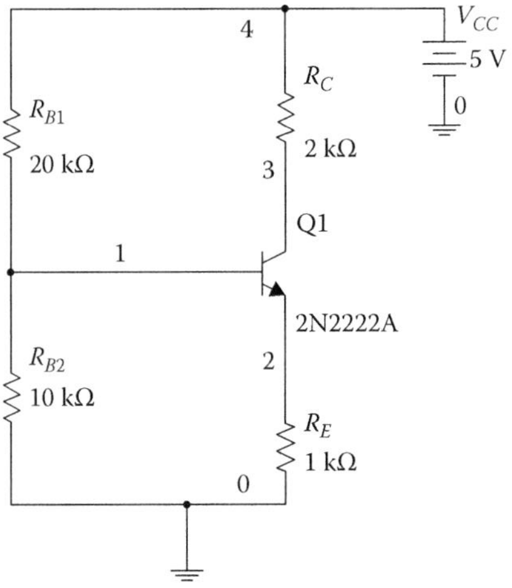

FIGURE P7.5
Circuit for determining the bias point of transistor 2N2222A.

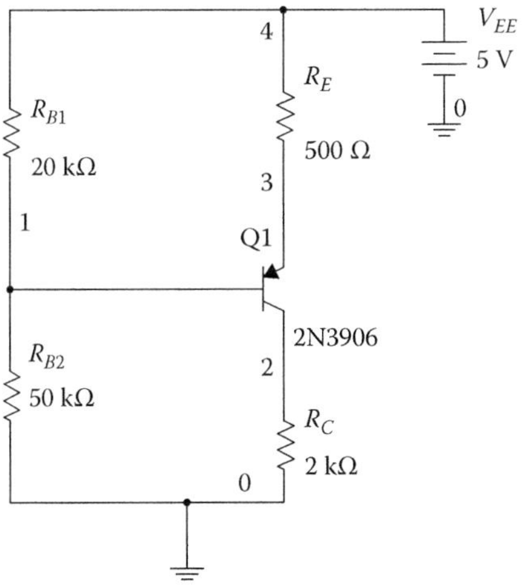

FIGURE P7.6
Circuit for obtaining the bias point of transistor 2N3906.

Problem 7.8 Use the Analog Discovery board to determine (1) the magnitude response of the circuit shown in Figure P7.8, (2) midband gain, (3) low cut-off frequency, (4) high cut-off frequency, and (5) bandwidth of the amplifier.

Problem 7.9 Use the Analog Discovery board to determine (1) the magnitude response of the circuit shown in Figure P7.9, (2) midband gain, (3) low cut-off frequency, (4) high cut-off frequency, and (5) bandwidth of the amplifier.

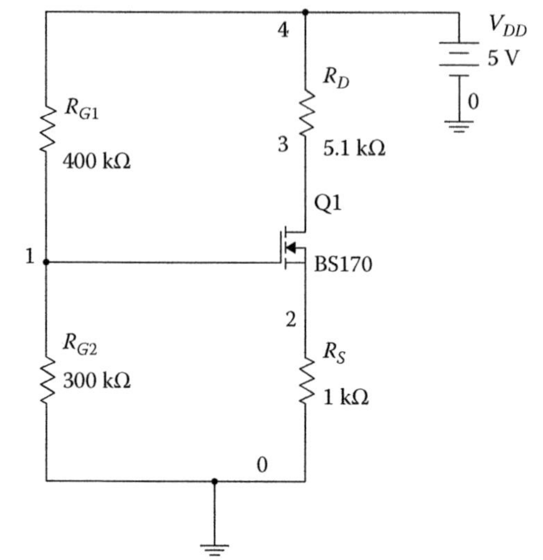

FIGURE P7.7
Circuit for obtaining the bias point of MOSFET BS170.

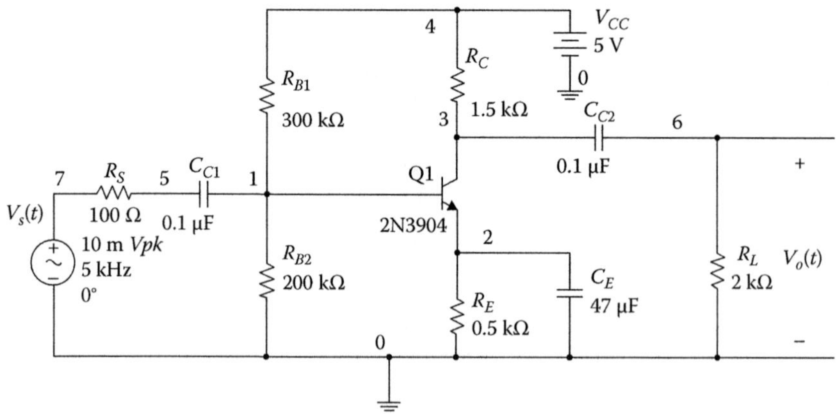

FIGURE P7.8
Common-emitter amplifier with transistor 2N3904.

Problem 7.10 For the circuit shown in Figure P7.10, use the Analog Discovery board (1) to determine gain at 4 kHz and (2) measure the voltage across the 100 Ω resistor and calculate the input resistance.

Problem 7.11 For the circuit shown in Figure P7.11, use the Analog Discovery board (1) to determine gain at 4 kHz and (2) measure the voltage across the 100 Ω resistor and calculate the input resistance.

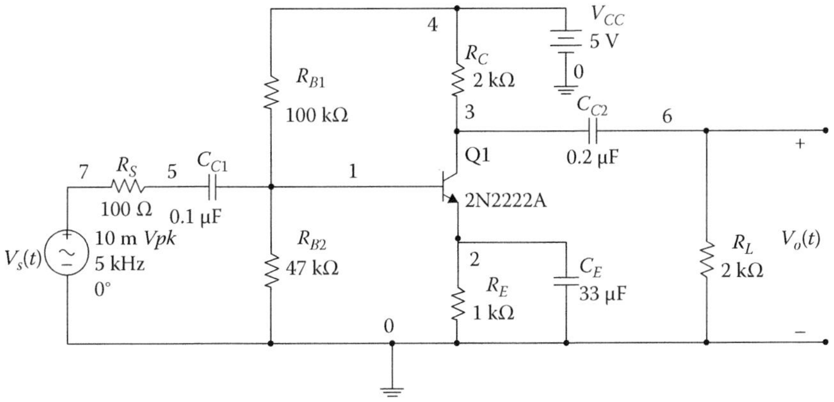

FIGURE P7.9
Common-emitter amplifier with transistor 2N2222A.

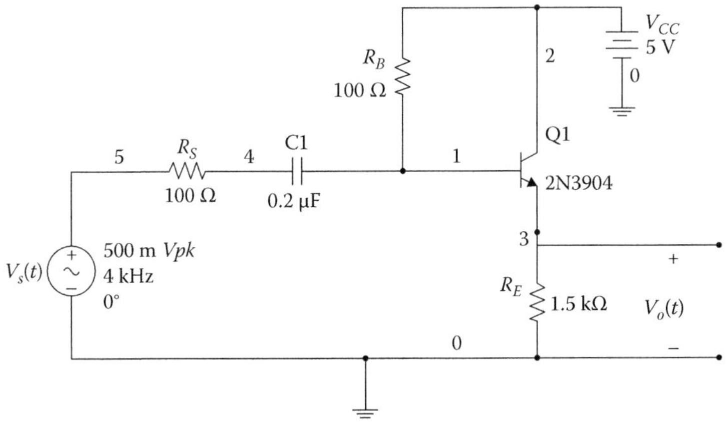

FIGURE P7.10
Common-collector amplifier with transistor 2N3904.

Problem 7.12 Use the Analog Discovery board to determine (1) the magnitude response of the circuit shown in Figure P7.12, (2) midband gain, (3) low cut-off frequency, (4) high cut-off frequency, and (v) bandwidth of the amplifier.

Problem 7.13 For the circuit shown in Figure P7.13, use the Analog Discovery board to find the voltage gain at 5 kHz. Determine (1) the magnitude response of the circuit, (2) low cut-off frequency, (3) high cut-off frequency, and (4) bandwidth of the amplifier.

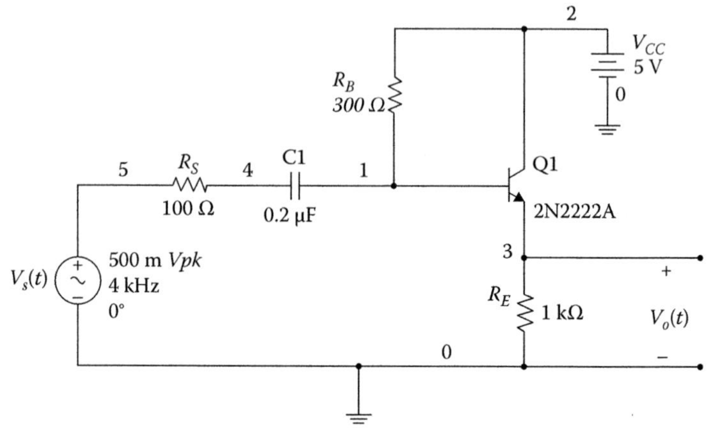

FIGURE P7.11
Common-collector amplifier with transistor 2N2222A.

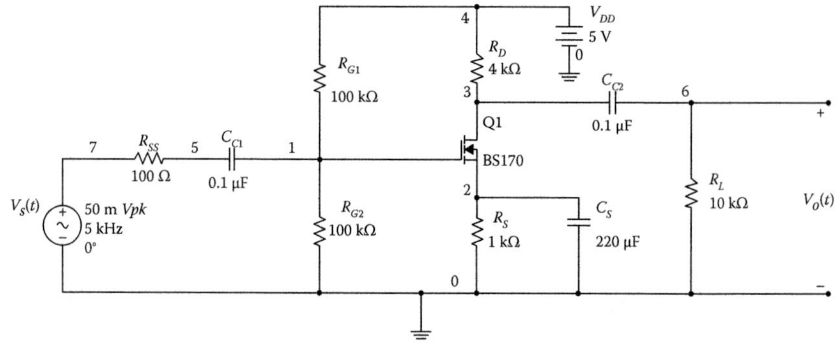

FIGURE P7.12
Common-source amplifier with transistor BS170.

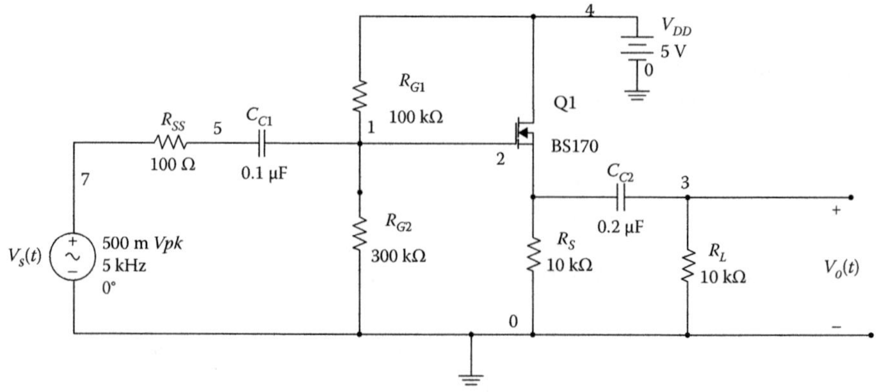

FIGURE P7.13
Common-drain amplifier with transistor BS170.

Problem 7.14 Use the Analog Discovery board to determine (1) the magnitude response of the circuit shown in Figure P7.14, (2) midband gain, (3) low cut-off frequency, (4) high cut-off frequency, and (5) bandwidth of the amplifier.

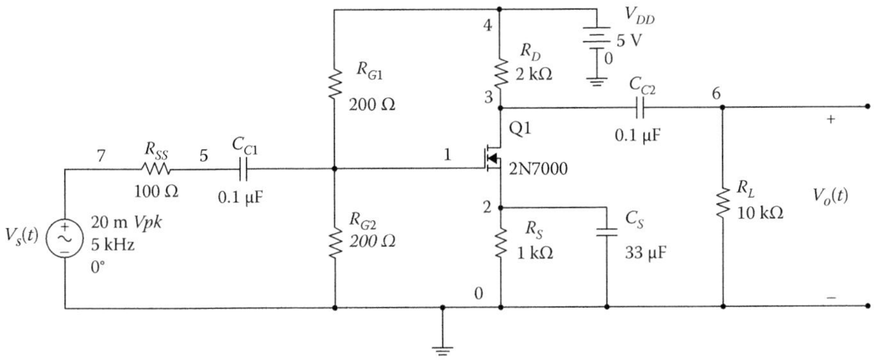

FIGURE P7.14
Common-source amplifier with transistor 2N7000.

Bibliography

Akhras, F.N. and J.A. Self, Modeling the process, not the product of learning, in Lajoie, S. P. (Ed.) *Computers as Cognitive Tools: No More Walls*, vol. 2, pp. 3–28, Lawrence Erlbaum Associates, Mahwah, NJ, 2000.

Alexander, C.K. and M.N.O. Sadiku, *Fundamentals of Electric Circuits*, 6th Edition, McGraw Hill, New York, 2016.

Analog Discovery Module, Information on the Analog Discovery module and supporting material. https://analogdiscovery.com/, June 19, 2017.

Attia, J.O. *Electronics and Circuit Analysis Using MATLAB*, 2nd Edition, CRC Press, Boca Raton, FL, 2004.

Attia, J.O. *PSPICE and MATLAB for Electronics: An Integrated Approach*, 2nd Edition, CRC Press, Boca Raton, FL, 2010.

Attia, J.O., L.D. Hobson, P.H. Obiomon, and M. Tembely, Engaging electrical and computer engineering freshman students with an electrical engineering practicum, *Proceedings of ASEE 12th Annual Conference and Exposition*, Columbus, OH, June 2017.

Attia, J.O., P. Obiomon, and M. Tembely, Hands-on learning for engaging freshman students through the electrical engineering practicum and the infinity project, *Proceedings of the 2016 ASEE-GSW Annual Conference* hosted by Texas Christian University, Fort Worth, TX, March 6–8, 2016.

Attia, J.O., M. Tembely, L. Hobson, and P. Obiomon, Hand-held mobile technology in a freshman course for enhanced learning, *Proceedings of the 2017 ASEE Gulf-Southwest Section Annual Conference*, organized by The University of Texas at Dallas, March 12–14, 2017.

Bowman, R. *Electrical Engineering Practicum*, Online Textbook, www.Trunity.com, 2014.

Bowman, R. Inspiring electrical engineering students through fully-engaged hands-on learning, *2013 IEEE 56th International Midwest Symposium on Circuits and Systems*, Columbus, OH, pp. 574–577, 2013.

Boylestad, R.L. *Introduction to Circuit Analysis*, 13th Edition, Pearson, Boston, 2015.

Brown, C.J., L.J. Hansen-Brown, and R. Conte, Engaging millennial college-age science and engineering students through experimental learning communities, *Journal of Applied Global Research*, vol. 4, pp. 41–58, 2011.

Byers, L.K., J.W. Kile, and C. Kiassat, Impact of hands-on first year course on student knowledge of and interest in engineering disciplines, *Proceedings of ASEE Annual Conference and Exposition*, Indianapolis, IN, June 2014.

Connor, K.A., B. Ferri, and K. Meehan, Models of mobile hands-on STEM education, *Proceedings of ASEE Annual Conference and Exposition*, Atlanta, GA, June 2013.

Connor, K.A., Y. Astatke, Dr. C.J. Kim, Dr. M.F. Chouikha, D. Newman, K.A. Gullie, A.A. Eldek, S.S. Devgan, A.R. Osareh, J.O. Attia, Dr. S. Zein-Sabatto, and Dr. D.L. Geddis, Experimental centric pedagogy in circuits and electronics courses at 13 universities, *Proceedings of ASEE 123rd Annual Conference and Exposition*, New Orleans, LA, June 26–29, 2016.

Connor, K.A, Dr. D. Newman, K.A. Gullie, Y. Astatke, M.F. Chouikha, C.J. Kim, Dr. O.E. Nare, J.O. Attia, Prof. P. Andrei, and L.D. Hobson, Experimental centric pedagogy in first-year engineering courses, *Proceedings of ASEE 123rd Annual Conference and Exposition*, New Orleans, LA, June 26–29, 2016.

Dori, H. and B. Breslow, How much have they retained? Making unseen concepts seen in a freshman electromagnetism course at MIT, *Journal of Science Education and Technology*, vol. 16, pp. 299–323, 2007.

Felder, R.M. and L. K. Silverman, Learning and teaching styles in engineering education, *Engineering Education*, vol. 78, no. 7, pp. 647–681, 1988.

Ferri, B., J. Auerbach, J. Michaels, and D. Williams, TESSAL: A program for incorporating experiments into lecture-based courses within the ECE curriculum, *ASEE Annual Conference and Exposition*, Vancouver, BC, June 2011.

Grout, I. and A. K. B. A'ain, Introductory laboratories in semiconductor devices using the Digilent Analog Discovery, *Proceedings of 2015 12th International Conference on Remote Engineering and Virtual Instrumentation (REV)*, Bangkok, Thailand, pp. 1–6, 2015.

Hendricks, R.W., K.-M. Lai, and J.B. Webb, Lab-in-a-box: Experiments in electronic circuits that support introductory courses for electrical and computer engineers, *Proceedings of ASEE Annual Meeting*, Portland, OR, June 12–15, 2005.

Henricks, R.W. and K. Meehan, *Lab-in-a-Box: Introductory Experiments in Electric Circuits*, 3rd Edition, John Wiley & Sons, Hoboken, NJ, 2009.

Holland, S.S., C.J. Prust, R.W. Kelnhofer, and J. Wierer, Effective utilization of the analog discovery board across upper-division electrical engineering courses, *Proceedings of 2016 ASEE Annual Conference and Exposition*, New Orleans, LA, June 2016.

Huette, L., Connecting theory and practice: Laboratory-based explorations of the NAE grand challenges, *Paper Presented at 2011 Annual Conference and Exposition*, Vancouver, BC, June 2011.

Irwin, J.D. and M. Nelms, *Basic Engineering Circuit Analysis*, John Wiley & Sons, New York, 2015.

Mazzaro, G.J. and R.J. Hayne, Instructional demos, in-class projects, and hands-on homework: Active learning for electrical engineering using the analog discovery, *Proceedings of 2016 ASEE Annual Conference and Exposition*, New Orleans, LA, June 2016.

Millard, D. Workshop—Improving student engagement and intuition with the mobile studio pedagogy, *Proceedings of the 38th ASEE/IEEE Frontiers in Education Conference*, Saratoga Springs, NY, pp. W3C-1, October 22–25, 2008.

Newman, D.L., G. Clure, M.M. Deyoe, and K.A. Conner, Using technology in a studio approach to learning: Results of a five year study of an innovative mobile teaching tool, in Keengwe, J. (Ed.) *Pedagogical Applications and Social Effects of Mobile Technology Integration*, Hershey, PA, 2013.

Newman, D., J. Lamendola, D.M. Morris, and K. Connor, Active learning, mentoring, and mobile technology: Meeting needs across levels in one place, in Keengwe, J. (Ed.) *Promoting Active Learning Through the Integration of Mobile and Ubiquitous Technologies*, Hershey, PA, 2015.

Nasr, K. J. and B.H. Ramadan, Impact assessment of problem-based learning in an engineering science course, *Journal of STEM Education*, vol. 9(3–4), pp. 16–24, 2008.

Prince, M. Does active learning work? A review of the research, *Journal of Engineering Education*, vol. 93, pp. 223–232, 2004.

Radu, M. Developing hands-on experiments to improve student learning via activities outside the classroom in engineering technology programs, *4th IEEE Integrated STEM Education Conference*, March 8, 2014.

Robertson, J.M., K. Meehan, R.J. Bowman, K.A. Connor, and D.A. Mercer, Exploiting a disruptive technology to actively engage students in the learning process, *Proceedings of the 2013 ASEE Annual Conference and Exposition*, June 2013.

Sedra, A.S. and Smith, K.C. *Microelectronic Circuits*, 6th Edition, Oxford University Press, New York, NY, 2010.

Yousuf, A., A. Wong, and D. Edens, Remote circuit design labs with analog discovery, *Proceedings of 2013 ASEE Annual Conference and Exposition*, Atlanta, GA, June 2013.

Zhang, L., I. Dabipi, Y. Jin, and P. Matin, Inspiring undergraduate students in engineering learning, comprehending and practicing by the use of analog discovery kits, *2015 IEEE Frontiers in Education Conference (FIE)*, El Paso, TX, pp. 1–4, 2015.

Index

C

D

U – V

W